D1305085

A CLEAR VIEW

BY JAMES CANNON
FOREWORD BY WILLIAM D. RUCKELSHAUS

GUIDE TO INDUSTRIAL POLLUTION CONTROL

RODALE PRESS, INC.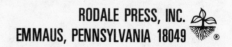
EMMAUS, PENNSYLVANIA 18049

Library of Congress Cataloging in Publication Data

Cannon, James Spencer, 1949-
  A clear view.

  "An Inform/Rodale Press book."
  Bibliography: p.
  Includes index.
  1. Environmental protection — United States —
Citizen participation. 2. Pollution — United States.
3. Factory and trade waste — United States.
I. Komanoff, Charles, joint author. II. Title.
TD171.C36   1976   363   76-3490
ISBN 0-87857-123-X
ISBN 0-87857-124-8 pbk.

An Inform/Rodale Press Book. Published by arrangement with Inform, Inc., 25 Broad St., New York, N.Y. 10004

Printed in the United States of America on recycled paper
First Printing — April, 1976
B-746

Foreword

It is difficult to find a public figure in America today who does not speak with great reverence about "participatory democracy," "openness in government," or "citizen involvement." This book seeks to give meaning to the call for citizen participation in the control and abatement of pollution. For this effort, and I think it is a successful one, the authors are to be greatly commended.

Regardless of one's position on the issue of how much openness in governmental decision-making processes is a good thing, or to what extent the citizen should participate in decisions, it is clear that if we once decide such openness and participation are necessary we should bend every effort to see that the participatory process we establish works.

In order for a citizen to meaningfully participate in government decision-making he or she must be informed and must have access to the decision-makers. This book imaginatively addresses both needs. It places great emphasis on the responsibility of a citizen to become knowledgeable prior to meaningful participation. It spells out how the citizen can become informed. Finally, it delineates how regulatory decisions are made and how they can be affected.

This alone would be a major contribution to making participatory democracy work in the pollution abatement field. But the book does more. It seeks to remove emotion from the processes we have established to control pollution. It recognizes that in debating an issue as complex as pollution it is not sufficient to discuss whether a standard is tough or weak, but focuses the citizen's attention on whether it is wise. The authors avoid endless polemic on whether the cause of pollution is a particular economic system. Instead, they attempt to steer the citizen to the real culprit, which he or she might affect, namely the individual plant. In short, the book seeks to prescribe how a modern American can give meaning to our historical tradition of citizen involvement, in a complex, difficult field.

If one of our country's major problems is a lack of citizen trust in our basic institutions, then surely one way of restoring trust is to insure that those institutions perform in a consistent and reliable manner. Involving citizens in an institutional action does two things toward this end. First, it insures a greater understanding by citizens of the complexity of the issues involved in any decision, and thereby enables them to better measure institutional performance. Second, it sensitizes the institution itself to the demands and expectations of the society in which it must function. It is only through efforts like this book that citizen involvement will advance from theory to practice, to the betterment of our society.

WILLIAM D. RUCKELSHAUS

Acknowledgements

The authors and editor would like to thank the many people who helped A CLEAR VIEW emerge from the darkness and acquire shape and substance. First and foremost, we must thank Joanna Underwood, INFORM's Director, for her contributions, enthusiasm, and unflagging attention to detail at every step of the project from conception to publication.

We are also extremely grateful to the many people who reviewed portions of the manuscript and assisted us greatly with both form and content. They include: Sandy Noyes, Marshall Beil, Edward Tuck, Anthony Wolff, and Robert Alexander, all members of INFORM's Steering Committee; George R. Webster, P.E., of Clayton Environmental Consultants; Lee Stephenson of Environmental Action; Richard Ayres of the Natural Resources Defense Council; Dr. Robert Harris of the Environmental Defense Fund; Swep Davis, at the U.S. Environmental Protection Agency; Prof. Raymond Bauer of the Harvard Business School; E. N. Brandt of Dow Chemical Co.; Kenneth Van Tine and H. N. Troy of Owens-Illinois; Paul V. Allemang, Gerald D. Rapp, Ralph H. Bernstein and Mark J. Thomas of the Mead Corp.; Bruce Marcus of the Financial Relations Board; William Reilly, Clem Rastatter, and Robert Cahn of the Conservation Foundation; and Robert H. Maynard of Smith and Schnacke.

For their help in research we would like to thank Mary Eyester, Ilene Steingut, Elizabeth Buckner and Pat Konecky. For assistance in copy-editing, we must thank Joyce Berry. Our typist-extraordinaire was Robert Szwed.

Finally, we greatly appreciate the generosity of the Mary Reynolds Babcock Foundation and the Abelard Foundation, who provided the financial support necessary for the project.

Contents

Smokestacks, industrial incinerator, Houston, Texas.

1

A Clear View:
A Tool for Change

A smokestack with a billowing black plume, for years the proud symbol of America's industrial wealth and technological prowess, has in the last decade acquired another meaning. As evidence has begun to accumulate of air pollution's role in causing disease, and as the country, increasingly urban, has begun to place a higher value on preserving the natural, the puffing smokestack has come to signify the Achilles heel, the unwanted by-products, of modern industrial society, rather than its strength. Companies sensitive to this change have replaced booming factories with rocks, rivers, and trees in their advertising copy and annual reports; smokestacks are shown devoid of plume, vestiges of bygone times.

Eliminating pollution from the environment has not proved as easy as eliminating it from brochures and magazines, however. Industry is now spending several billion dollars a year on pollution control, and is expected to have to spend at an even higher annual rate for the next ten years to meet legal standards. Some companies have made real progress in cleaning up heavily polluting plants. But at last count—1973—there had been little change in overall national air pollution levels (though in this case holding the line might be considered an advance). Industrial plants were still a major part of the problem (see box). A basis for altering the way we use the natural environment has been established. But while the groundwork for cleaning up industrial and other kinds of pollution has been laid, the actual effort has barely begun.

In 1970, Congress passed the Clean Air Act, the first comprehensive legislation to reduce air pollution in the United States. This was complemented in 1972 by the similarly aimed Water Pollution Control Act. Both dealt with industrial sources of pollution—whose high visibility had made them the focus of much public attention—as well as with other major sources of air and water pollution, most notably automobiles and municipal sewage plants.

The ensuing monumental effort to establish both the reams of regulations and the layers of bureaucracy to carry out these new laws, in an area which involves not only a new national concern but much new science and technology, has, not surprisingly, come up against serious obstacles. Companies and regulators have gone

NATIONAL AIR POLLUTION LEVELS, 1973

(million tons per year)

Source of Emissions	Particulates	Sulfur Oxides	Nitrogen Oxides	Carbon Monoxide	Hydro-carbons
Transportation	1.3	.8	11.0	79.3	13.7
Fuel Combustion in Stationary Sources (primarily electric utilities, factories)	6.4	25.6	11.2	1.0	1.7
Industrial Processes	11.9	6.7	0.6	13.0	3.1
Solid Waste	.6	—	0.1	2.8	0.7
Miscellaneous	.8	.1	0.1	4.8	12.1
TOTAL	21.0	33.2	23.0	100.9	31.3

Source: U.S. Environmental Protection Agency, National Emissions Data Bank

to court in time-consuming disagreement over policy questions, such as whether very tall smokestacks—hundreds of feet high—which disperse rather than capture pollutants, constitute an acceptable method of achieving the aims of the Clean Air Act. Industry and environmentalists have also challenged specific regulations in court—either for being too strict, or too lenient—including every U.S. Environmental Protection Agency (EPA) guideline for levels of water pollution control which various industries must achieve by 1977 and 1983. Many state regulatory programs are not proceeding on schedule. Entire areas of the law—such as limits for release of especially toxic chemicals, and regulation of water pollution from "non-point" sources such as construction sites—have been largely ignored while regulators and industry alike grapple with cleaning up the "major" pollutants from the "major" polluters.

The push to improve environmental quality has also produced problems that the new statutes did not anticipate. We must face economic trade-offs. For example, what should be done when forcing an old, inefficient, but still job-generating factory to spend several million dollars for pollution controls will cause it to become unprofitable and close? We must understand and cope with environmental trade-offs, too. For instance, is it best to control sulfur dioxide emissions from electric power plants by removing sulfur from smokestack gases (creating a potential large sludge residue disposal problem), by burning low-sulfur fuels (possibly increasing pressure to strip-mine coal in the Great Plains), or by encouraging development of nuclear power (causing problems in safety and radioactive waste disposal)?

The newness and complexity of many of these issues, and the fact that laws are on the books, has produced a tendency to leave many crucial decisions to the "experts." Yet this was not the intent of the 1970 and 1972 laws, nor is it in the nation's democratic tradition. One of the reasons that implementation of the Clean Air Act was organized on a state-by-state basis was to allow greater local involvement in establishing regulations. The Water Pollution Control Act specifically called for public participation in granting discharge permits, developing area-wide waste management programs, and making EPA rules. Both acts made a point of requiring that the public have access to the statistical air and water emissions data collected by government agencies, a necessity for public input on decisions. In fact, the very complexity of trade-offs between kinds and degrees of environmental controls, and among competing considerations of economics, energy, resources, and land use, requires more—not less—public participation in decision-making. These are not purely technical problems, but rather far broader questions, regarding the needs, desires, and values of the whole society.

A CLEAR VIEW offers the concerned citizen the tools needed to be an effective and competent participant in decisions about pollution control at an existing or a proposed new factory. At first thought, the complex industrial technical systems involved in factory operations may seem formidable and forbidding. A layperson starting from scratch will have to put in a good deal of time and concentrated effort to unravel the workings of a manufacturing process, its pollution problems, the solutions available and their costs. Nevertheless, using the methods in A CLEAR VIEW citizens can come to relatively sophisticated conclusions about a factory's environmental impact.

Federal environmental laws have largely insured that the basic data necessary for public participation in local factory pollution control decisions—whether formal, as in testimony at regulatory hearings, or informal, as in discussions with company managers—exist and are available. The chapters that follow set out how and where to obtain this information and describe a systematic approach to analyzing and applying it. The approach described can be used by any citizen organization, college student group, or union—the "you" to whom this book is addressed. It may also be useful to investors who wish to know whether a company's pollution problems are likely to provoke a lawsuit or cause damage to its public image; or to corporate managers seeking a new perspective on and evaluation of factory pollution control problems.

Chapter 2, "Understanding the Problem," describes how to learn about a company, about a production process and its potential to pollute, about the effects of such pollution, and about possible control methods and their costs—a process which is a prerequisite to a valid evaluation of a factory's actual air and water emissions. While it is not imperative to do immediately all the research outlined, it is essential to acquire some background in each of these areas, before proceeding to gather specific factory pollution data. (A comprehensive list of all books and other references mentioned appears, along with full information on how to obtain them, at the back of the book.)

Chapter 3, "Gathering Factory Data," describes the specific facts and figures necessary for an evaluation of a factory's pollution control problems: air and water emissions statistics, percent efficiencies of existing pollution control devices, capacity, age and output of production equipment, raw material and energy use data, expenditures for pollution control, and legal status. And the chapter discusses how to extract the information from the relevant sources: state air and water pollution control agencies, regional EPA offices, the courts, and the company operating the factory.

Chapter 4, "Evaluating the Data," is about what to do with the information once it has been acquired. The chapter illustrates how all the important facts can be organized into thirteen charts. Each of these answers an important question about pollution control at the factory. The chapter describes how to use these charts to evaluate the factory's pollution record in relation to four "yardsticks" of performance. These measures are: legal standards, pollution levels causing adverse effects on the environment, levels achievable using the best available (generally known as "state-of-the-art") controls, and the performance of other factories in the industry. Finally, the chapter discusses how to draw conclusions about the factory—whether it is clean or dirty, doing the best it can, or making little effort—and how to develop proposals for improvement.

Chapter 5, "Having an Impact," discusses ways in which citizens can use their conclusions to participate in pollution control decisions about the factory—talking to company managers, participating in the regulatory process, and presenting proposals to the public.

Chapter 6, "New Plants: Preventive Medicine," deals with the special regulations and sources of information, and the unique problems and opportunities, associated with preventing environmental problems at a new industrial facility. It is intended to be used in conjunction with the material presented in previous chapters.

Clearly, neither every source of information described nor every method of analysis will be applicable to every factory. People will also vary in the amount of time they can devote to understanding factory conditions and in what they hope to accomplish from such an inquiry. Thus, while A CLEAR VIEW tries to present some of the potentially more complicated and confusing aspects of analyzing industrial pollution control in a clear, step-by-step form, and to describe the full range of information sources available, it should not be considered a rigid prescription. Each individual should use those portions which answer his or her needs and adapt them to the situation at hand.

This book's intent is to provide anyone having energy, a certain seriousness of resolve, a grasp of basic math, and the use of a pocket calculator, the tools with which to participate effectively in decisions heretofore left largely to technicians. Needless to say, these are decisions which will significantly determine the kind of world in which we and those who follow us will live.

2

Understanding the Problem

Recognizing most industrial pollution isn't difficult: a black plume in the sky, a bad smell in the air, an oil slick across a river below a discharge pipe. Determining what can be done about it is harder: how much can the pollution be reduced? by what methods? what would it cost?

If you wish to understand what can be done about a particular factory's pollution problems, it is necessary to do some background research into how the plant works, available control methods, and factors affecting the pollution control policies of the company which operates it.

This chapter describes how to do such research. The references discussed should allow you to acquire a basic understanding of: the company, industry and factory production technology (including factory processes, raw materials and energy use), the factory's potential to pollute, the environmental effects of pollution, available control methods, and the economic and social factors affecting a company's pollution control decisions. You can then go on to gather and evaluate specific factory air and water pollution emissions data as described in Chapters 3 and 4, although you may want to return to some of the sources of information noted in this chapter, as you proceed.*

---

* All references noted appear, grouped by chapter, with full bibliographical data and information on how they can be obtained, in the "References" section at the back of this book.

THE COMPANY

Your first step should be to become acquainted with the company owning and operating the factory with which you are concerned, since the company is responsible for purchasing and utilizing pollution controls. Begin with the company's annual report. Any corporation whose stock is traded on the New York Stock Exchange is required to publish an annual report at least fifteen days before its annual meeting. Corporations generally will send a free copy to anyone upon request. Annual reports usually include a great deal of background information on the company's major industrial operations, the products made at these plants, and subsidiary firms, and a five- or ten-year summary of financial data. In reading annual reports (as well as other background information sources, such as SEC filings, magazines, newspapers, and trade publications discussed below), you should pay particular attention to any information about the size and production of the factory you are interested in, the location of similar factories, the amount of money invested to date at various locations for pollution control, and the type of pollution controls installed and planned.

Securities and Exchange Commission (SEC) requirements have created additional worthwhile sources of information. Companies must submit detailed reports on their operations to the SEC. The submissions, called "10K" (filed annually) and "8K" (filed monthly) forms, are available for inspection at the Washington and regional offices of the SEC (see list of addresses), and in some cases can be ordered from the company (price range 0—$7.00; see annual report).

Whenever a company makes a new stock offering or floats a bond issue, the brokerage house which sells the issue is required by the SEC to distribute a prospectus describing the firm's operations and financial position. Lawsuits, and pollution problems that might "materially affect" company earnings must be included. If you notice in a company annual report or in the newspapers that a stock or bond issue has been made in recent years, obtain a copy of the prospectus from any of the brokerage houses handling sales of the issue, or from the company. These documents are usually free, and are available upon request as long as the supply lasts.

SECURITIES AND EXCHANGE COMMISSION REGIONAL OFFICES

BOSTON
150 Causeway Street
Boston, Massachusetts 02114

NEW YORK
26 Federal Plaza
New York, N. Y. 10007

ATLANTA
1371 Peachtree Street
NE Suite 138
Atlanta, Georgia 30309

CHICAGO
219 South Dearborn Street
Room 1708
Chicago, Illinois 60604

FORT WORTH
503 United States Courthouse
10th and Lamar Streets
Fort Worth, Texas 76102

DENVER
2 Park Central
Room 640
1515 Arapahoe Street
Denver, Colorado 80202

LOS ANGELES
312 North Spring Street
Room 1043
Los Angeles, California 90012

SEATTLE
3040 Federal Building
915 Second Avenue
Seattle, Washington 98174

WASHINGTON, D.C.
4015 Wilson Boulevard
Arlington, Virginia 22203

Interesting and important information about the factory and company can also be gleaned from general readership magazines, such as *Time* (Chicago, $18 a year), *Newsweek* (New York, N. Y., $19.50 a year) and *Business Week* (New York, N. Y., $17 a year). To locate such articles, consult the *Reader's Guide to Periodical Literature*, an annual index (by subject) of articles appearing in nearly 200 magazines. Informative articles about the company have probably also appeared in *The New York Times* and the *Wall Street Journal*. Both newspapers also publish an annual index. All three indexes should be available in general reference libraries.

Another good potential source of information is trade organizations, such as the American Paper Institute, the American Petroleum Institute, and the International Lead and Zinc Research Institute, organized to collect and disseminate information and to lobby for an industry. *The Encyclopedia of Associations, Vol. 1: National Organizations in the U. S.* (Gale Research Co.,

Detroit, Michigan, $55.00) available in most business libraries, lists the names and addresses of thousands of these groups. You should note those for the industry to which the factory you are studying belongs, and write to them requesting any free descriptive literature relating to industry pollution control efforts.

Trade organizations, as well as independent publishers such as McGraw-Hill, publish journals known as "trade magazines," which are also available in most business libraries. These journals report news within a particular industry, and include somewhat more technical articles about the operations of member companies. The articles in these journals are intended for, and are often written by, individuals connected with firms in the industry. By reviewing back issues a layperson can frequently learn a great deal about a company's activities in recent years and can become familiar with the technical language of the industry.

Three indexes catalogue articles appearing in trade and business magazines. The *Funk and Scott Index of Companies and Industries* lists articles from 750 magazines. Articles in *Funk and Scott* are indexed numerically, according to the SIC numbers (described below) of the industry they discuss, and alphabetically, by company. The annual *Applied Science and Technology Index* catalogues articles from 220 magazines, and the *Business Periodical Index* catalogues 150. The latter two include some publications not in *Funk and Scott*. All three can be found in business libraries.

INDUSTRY AND FACTORY PRODUCTION

Once you have done some general reading on the company and factory, you can begin focusing on specifics. First, you should determine exactly the type of industrial operation you are dealing with. While in most cases this is obvious—a paper mill produces paper products and the operating company is a member of the paper industry—in some cases it is not so clear. The largest coke (a coal product) producing plant in the world is operated by the

*Paper mill, operated by International Paper at Corinth, N. Y.,
on the Hudson River. Water pollution treatment facilities are
at upper left.*

U. S. Steel Corporation, not a coke company, for example, and in fact, most coke is produced by steel companies. Descriptions of coke technology, specialized pollution control systems for coke ovens, and pollution regulations applicable to coke ovens can best be obtained from documents and specialists in the steel industry.

The Federal government has developed a system to classify every industrial process by the industry to which it belongs. The system, called the Standard Industrial Classification, assigns a four-digit number to each industrial operation (see chart of main classifications). The first two digits are the same for each major industrial group; the last two indicate a particular type of operation within that group. For example, all aspects of paper processing have 26 as the first two digits of their SIC code; pulp mills are SIC 2611, and paperboard mills are 2631.

You can determine the SIC code for the factory you are interested in by looking up the name of its industrial process, or of the product it produces, in the *Standard Industrial Classification Manual*, available from the U. S. Government Printing Office, Washington, D. C. (GPO # 4101-0066, $6.75), and at most business libraries. Knowing a factory's SIC code is helpful in placing it in context with the rest of the industry, and is often useful in obtaining information about it.

Industry Production Methods

Having determined the industrial category to which the factory you are studying belongs, you should turn attention next to learning how that industry makes its product—produces paper, makes steel, etc. There are often several ways a particular product can be manufactured. You should be aware of each method, since each generally produces different kinds and amounts of pollution. In fact, switching or adapting a factory's production technology is sometimes the most feasible method of pollution control.

Trade organizations usually publish "how to" booklets which explain their industries' technologies in very simple terms. If such a booklet is not available for the industry with which you are concerned, the *Industrial Pollution Control Handbook*, Herbert Lund, editor, (McGraw-Hill, New York, $29.50) and other books avail-

---

STANDARD INDUSTRIAL CLASSIFICATIONS
MAIN MANUFACTURING GROUPS

Food and kindred products	2000
Tobacco manufacturers	2100
Textile mill products	2200
Apparel and other fabric products	2300
Lumber and wood products	2400
Furniture and fixtures	2500
Paper and allied products	2600
Printing and publishing	2700
Chemicals and allied products	2800
Petroleum refining	2900
Rubber and plastics	3000
Leather and leather products	3100
Stone, clay, glass and concrete	3200
Primary metals industries	3300
Fabricated metal products	3400
Machinery, except electrical	3500
Electric and electronic machinery	3600
Transportation equipment	3700
Analytical, photographic, medical, optical instruments; watches and clocks	3800
Miscellaneous manufacturing industries	3900

---

able in most engineering and business libraries describe the technology and pollution problems associated with various industries.

One other major source of information on an industry's production technology may be available. In conjunction with promulgation of the water pollution standards mandated by the 1972 Federal Water Pollution Control Act, the U.S. Environmental Protection Agency (EPA) has published detailed studies of each of various industries' water pollution problems, called *Development Documents for Effluent Limitation Guidelines and New Source Performance Standards*. Thirty-three *Development Documents* have appeared so far, and are available from the U.S. Government Printing Office (GPO # varies, $1.60–$20.30, see list). If one of these

DEVELOPMENT DOCUMENTS
FOR EFFLUENT LIMITATIONS GUIDELINES AND
NEW SOURCE PERFORMANCE STANDARDS

Industry	GPO Order No.	Price
Pulp and Paper—Unbleached Kraft and Semi-chemical Pulp	5501-00910	$3.45
Builders Paper and Roofing Felt	5501-00909	$1.75
Red Meat Processing	5501-00843	$2.20
Dairy Product Processing	5501-00898	$2.05
Grain Processing	5501-00844	$1.75
Citrus, Apple and Potato Processing	5501-00790	$2.45
Catfish, Crab, Shrimp & Tuna Processing	5501-00920	$4.50
Beet Sugar	5500-00117	$2.00
Cane Sugar Refining	5501-00826	$2.10
Textile Mills	5501-00903	$2.65
Cement Manufacturing	5501-00866	$1.60
Feedlots	5501-00842	$3.25
Electroplating—Copper, Nickel, Chrome & Zinc	5501-00816	$2.40
Organic Chemicals—Major Organic Products	5501-00812	$3.60
Inorganic Chemicals—Major Inorganic Products	5502-00121	$3.60
Plastics—Synthetic Resins	5501-00815	$2.65
Soap & Detergent Manufacturing	5501-00867	$2.35
Basic Fertilizer Chemicals	5501-00868	$2.00
Petroleum Refining	5501-00912	$2.75
Steel Making	5501-00906	$20.30
Bauxite Refining	5500-00118	$1.45
Primary Aluminum Smelting	5501-00817	$1.80

Industry	GPO Order No.	Price
Secondary Aluminum Smelting	5501-00819	$1.70
Phosphorous Derived Chemicals	5503-00078	$1.90
Steam Electric Powerplants	n.a.	$8.90
Power Plants—Cooling Water Intake Structure Technology	n.a.	n.a.
Ferroalloys—Smelting and Slag Processing	5501-00780	$2.10
Leather Tanning and Finishing	5501-00818	$1.95
Insulation Fiberglass	5501-00781	$1.50
Flat Glass	5501-00814	$1.65
Asbestos—Building, Construction & Paper	5501-00827	$1.70
Rubber Processing—Tire & Synthetic	5501-00885	$2.25
Timber Products—Plywood, Hardboard & Wood Preserving	5501-00853	$3.30
	EPA Order No.	
Meat Rendering*	EPA 440/1-74/031D	$2.70
Formulated Fertilizer*	EPA 440/1-74/042A	$1.40
Asbestos—Textiles, Friction Material & Seeding Devices*	EPA 440/1-74/035A	$1.70
Grain Mills—Animal Feeds, Breakfast, Cereals & Wheat Starch*	EPA 440/1-74/039A	$1.90
Non-ferrous Metals Manufacturing—Copper, Lead & Zinc*	EPA 440/1-75/032	n.a.

---

* As of early 1975 these documents were available only through the EPA. They will be available from the GPO in the future. Refer to the EPA number in obtaining the future GPO number from the GPO.

n.a. signifies not available as of early 1975.

---

reports has been prepared for the industry you are studying, obtain it. It should contain a thorough discussion of the industry's production technology, as well as a wealth of useful information on the industry's water pollution problems.

A helpful way to check your understanding of industry production technology is to try drawing a simple "flow diagram" of typical factories in the industry, graphically tracing the various production steps. For example, in the steel industry, coal is converted into coke in coke ovens, coke is mixed with iron ore and limestone in blast furnaces to produce pig iron, pig iron is refined into steel in steelmaking furnaces, and raw steel is molded into a wide range of products in rolling and finishing mills.

Factory Equipment

Having surveyed the production processes used in the industry, you next need to decipher the exact type and size of production equipment operating at the particular plant you are investigating. Each furnace, oven, mill and boiler has a different potential to pollute, and pollution regulations are often set on a process-by-process basis. It is extremely helpful to develop a "process profile" for the factory, similar to the industry flow chart, on which the actual production equipment operating, the capacity of each piece of equipment, and its actual output for the preceding year, if available, are noted.

Most trade organizations publish a catalogue of major factories within their industries which includes the type and capacity of the production equipment present at each. The American Iron and Steel Institute publishes the *Iron and Steel Works Directory* every other year, for example, which lists furnaces and ovens at steel mills. *Lockwood's Directory* lists the equipment at the pulp and paper plants in this country. Many of these directories can be found in business libraries or at the respective trade organization headquarters.

Equipment listed in a directory may have been retired, or be on stand-by service. It is important, therefore, to verify the operating status of each production unit with a person familiar with the plant—a plant employee, a company official, or an engineer at the local pollution control agency. (How to approach the company for information is fully discussed under "Company Sources" in Chapter 3.)

Factory Raw Materials Use

You should also know the kinds and, if possible, amounts of raw materials the factory uses in its production processes, as this is often another important key to understanding its environmental problems. For example, copper is produced from a sulfur-bearing ore; copper smelters thus have a large sulfur pollution problem to contend with. In fact, air pollution regulations for copper smelters are often expressed in terms of the percent of sulfur in factory raw materials which must be kept from escaping up the smokestack.

Eventually, you will also want to pay particular attention to whether factory raw materials contain any especially "toxic" substances, most notably heavy metals such as arsenic, mercury, cadmium and lead, or certain organic chemicals. If they do, you should be alert in your research to indications that even very small amounts of the toxic substances are being emitted into air or water.

Raw material figures are reported on "National Pollution Discharge Elimination System" water discharge permit applications and on some "Refuse Act permit" applications. One or the other of these applications will have been filed for the factory you are concerned about, at your regional EPA office. Obtaining and using this extremely useful document is described fully in Chapter 3, under "Water Pollution Data."

If the permit application omits raw material use data, then you can try to find this information in industry directories. It is also possible at this stage to get at least a rough idea of raw material use from industry-wide totals reported annually by most trade associations. The American Iron and Steel Institute, for example, publishes an *Annual Statistical Report* which lists total raw materials and energy consumption for the entire steel industry. Multiplying industry-wide totals by the percentage of

18

Raw materials for copper smelters: copper mine near Tucson, Arizona.

Energy use: transmission lines into Bethlehem Steel mill, Sparrows Point, Maryland.

total industry production accounted for by the factory yields its approximate materials demand.

One other very useful reference for industry-wide raw material, fuel, and energy use figures (and production, capital expenditure, employment and payroll data as well) is the *1972 Census of Manufacturers*, compiled by the Bureau of the Census, U.S. Dept. of Commerce. The entire *Census* is available in many libraries. Portions of it relating to various industries can be ordered directly from the Government Printing Office (request Publications Order Form No. 45 from the U.S. Dept. of Commerce to obtain the order number for the industry segment you want; price range $.75–$1.40).

Factory Energy Use

All factories use energy, in addition to the basic raw materials of production, to make their products. This energy may be produced by burning fuels at the factory, or it may be purchased, in the form of electricity, from a utility. Wherever it is produced, however, its generation entails serious environmental problems, ranging from air pollution from coal-combustion to production of radioactive wastes by nuclear reactions. Understanding energy use is thus important in analyzing a factory's total environmental impact.

Try to determine specifically how the factory you are investigating meets its energy needs. First, most plants burn fossil fuels (coal, oil, or natural gas), to produce steam or hot air for industrial processes, to run motors and assembly lines, to heat furnaces requiring high temperatures, or simply to heat the workplace. The sources which provide data on raw material use, described above, should include figures on the factory's consumption of coal, oil and gas for these purposes.

The factory may also purchase electricity to run its operations: a small amount, designed to meet only basic needs like lighting, or a very large amount, designed to be used in production. A medium-sized aluminum smelter, which uses electricity to turn alumina ore into pure metal, can consume as many kilowatt-hours annually as 200,000 all-electric homes. Information on the amount of electricity, measured in kilowatt-hours, which a factory purchases each year is often available from the electric utility supplier, or from the state Public Service Commission.

A number of studies have analyzed the energy use of various industries, and assessed factors which may have caused consumption to increase or decrease in recent years. A Ford Foundation-sponsored study, called *Potential Fuel Effectiveness in Industry* (Ballinger Publishing Co., Cambridge, Mass., $2.50) describes in clear terms fuel and energy use patterns in the iron and steel, petroleum, paper, aluminum, copper and cement industries. A study by the Conference Board, a New York, industry-supported research organization, entitled *Energy Consumption in Manufacturing* (Ballinger Publishing Co., Cambridge, Mass., $9.95), provides similar but more detailed data on all the above industries (except copper), as well as on the food, chemicals and glass industries. Either of these studies may help you understand how and why the factory you are studying uses energy.

POLLUTION POTENTIAL

After learning about the company, and about industry and factory production technology, it is useful to estimate the nature and magnitude of problems the factory you are studying might have in the areas of air and water emissions, release of toxic substances, and solid wastes, *i.e.*, its "potential to pollute." Your research and data-gathering efforts can then be concentrated on understanding the possible effects of factory pollutants on the environment, and how the company could be and is controlling them.

With the help of EPA data, the factory's potential air and water pollution discharges can be estimated quite specifically. For many major industries, the EPA has in recent years calculated what it calls air pollution "emission factors" and water pollution "raw waste loads." These numbers indicate the amount of pollution anticipated from an industrial process operating without benefit of pollution controls. These factors are usually expressed in pounds of pollutant which would be released per ton of product produced.

Catalytic cracking unit, oil refinery, Joliet, Illinois. Catalytic crackers have a high air pollution potential.

Potential Air Pollution

The EPA has established air pollution "emission factors" for literally hundreds of industrial processes, using engineering estimates and actual tests on the air escaping from plant equipment. They appear in an EPA book entitled *Compilation of Air Pollutant Emission Factors, Second Edition*, (GPO # EP. 4.9 42/2, $2.90) which can be ordered from the Government Printing Office, Washington, D. C. Each of the ninety industry sections in the book includes a chart listing the "emissions factors," or pollution potential, for each production process in the industry, and, in addition, a description of major raw materials and products, basic technology of production, sources of each major air pollutant, and available pollution control methods.

In the paper industry, for example, eight industrial processes are analyzed for production of five common air pollutants (the portion of the chart for particulates is reproduced opposite). Emissions from processes operating with some commonly-used air pollution control systems are also included.

The burning of fuels in power boilers is dealt with separately in the *Compilation*. For each type of fuel (gas, oil, coal), emission factors, in pounds of pollutant per unit of fuel burned, for various types of boilers are listed. (Power boiler emission factors thus differ from industrial process factors in that they are expressed per unit of raw material used rather than per unit of product produced.)

For example, under anthracite coal combustion, three types of boilers are analyzed for six air pollutants. For particulate and sulfur dioxide emissions, the chart indicates that the number given must be multiplied by the percent of ash or sulfur in the coal to obtain the emissions released per ton of coal (see particulate portion of chart reproduced opposite).

It is possible to estimate the total potential emissions of any major pollutant from the factory you are concerned with, using EPA air emission factors, if you know the capacity of the production processes employed, and the factory's fuel use. (Obtaining these figures is described under "Factory Equipment" and

EMISSION FACTORS FOR SULFATE PULPING
(unit weights of air-dried unbleached pulp)

SOURCE	TYPE OF CONTROL	PARTICULATES (lbs./ton pulp)
Blow tank accumulator	Untreated	—
Washers and screens	Untreated	—
Multiple-effect evaporators	Untreated	—
Recovery boilers and direct-contact evaporators	Untreated	151
	Electrostatic precipitator	15
	Venturi scrubber	47
Smelt dissolving tank	Untreated	2
Lime kiln	Untreated	45
	Scrubber	4
Turpentine condenser	Untreated	—
Fluidized-bed calciner*	Untreated	72
	Scrubber	0.7

* Only a few plants in the western U.S. use this process

EMISSIONS FROM ANTHRACITE COAL COMBUSTION WITHOUT CONTROL EQUIPMENT

TYPE OF FURNACE	PARTICULATES (lbs./ton coal)
Pulverized (dry bottom), no fly-ash reinjection	17A*
Overfeed stokers, no fly-ash reinjection	2A*
Hand-fired units	10

* A is the ash content expressed as weight percent

"Energy Use" above.) First multiply the capacity of each piece of production equipment (expressed in tons/day) by the EPA emission factors (in lbs/ton of product) for the pollutant. Add the results to obtain the potential daily emissions, in pounds, of that pollutant from factory production. For example, using the emission factors for particulates presented in the tables on page 23, if a pulp mill produces 1,000 tons of pulp a day, using recovery furnaces and a lime kiln, the plant will generate approximately 151,000 (151 x 1,000) lbs. of particulates a day at the furnaces, and 45,000 (45 x 1,000) lbs. a day at the kiln. Particulates generated at both production processes total 196,000 lbs. a day.

Then multiply the amount of fuel burned (in tons, thousand gallons, or trillion cubic feet per day, depending on whether you are dealing with coal, oil or gas) by the appropriate emission factor for the same pollutant. If the mill in the example above burned 250 tons a day of anthracite coal containing 10 percent "ash" (*i.e.* non-combustible minerals) at "overfeed stoker" power boilers, then, according to the factors listed in the chart, the particulate load there would amount to 5,000 (2 x 10 x 250) lbs. a day.

This, added to the previous total, equals the approximate potential daily emissions, in pounds, of the pollutant from the entire factory.* Thus particulate pollution generated for the entire pulp mill in the above example would be 201,000 lbs. a day.

Remember, this amount reflects only *potential* plant air pollution, not real emissions (unless, of course, the plant has no pollution controls whatsoever).

Potential Water Pollution

The 1972 Water Pollution Control Act ordered the EPA to similarly estimate the water pollution generated by industrial processes operating without any water treatment equipment. For some in-

---

* A shortcoming of this estimate is that it includes potential pollution from only the major processes. No emission factors have been calculated for many minor plant operations or for sources of "fugitive" emissions such as dust from raw material stockpiles.

dustries, emission factors, called "standard raw waste loads," were incorporated into the *Development Documents*. They also appear in other technical sources on industrial pollution, including the McGraw-Hill *Industrial Pollution Control Handbook*, mentioned above.

"Standard raw waste loads" differ from air emission factors, however, in that they are usually indicated in terms of a range of discharges. This is because the amount of pollution getting into waste water can vary greatly from plant to plant, even where the plants employ the same production process, due to differing operating procedures.

Thus the *Development Documents* usually present raw waste loads for a number of plants of various sizes. The loads are reported in two ways: in terms of milligrams of pollutant in an average liter* of waste water, and in terms of pounds of pollutant released per ton of product produced.

Try to find sample raw waste loads for one or more factories approximately the same size, in terms of production, as the plant you are studying. If you take their raw waste loads, expressed in terms of pounds of pollutant per ton of product, and multiply them by your factory's production in tons per day, you will have a good idea of the pounds of various pollutants which the factory might be emitting daily.

Potential Toxic Substance Release

The *Compilation* and *Development Documents* describe only potential emissions of major industry pollutants. However, additional pollutants, including highly toxic ones, such as hydrogen sulfide and arsenic, may also be released into the air or water at the factory in lesser quantities. Relatively minute amounts of these chemicals can cause severe damage in a short space of time (a dose of .0003 pounds of arsenic, for example, is fatal to humans).

---

* A milligram of pollutant in a liter of water (mg/1) is approximately equivalent to one "part" or unit of pollutant per million parts of water (ppm). The two terms, *milligrams per liter* and *parts per million*, are often used interchangeably in discussions of water pollution.

The EPA has authority under existing air and water pollution control acts to regulate emissions of toxic pollutants. However, the agency had done little in this area, anticipating Congressional passage of a separate Toxic Substance Control Act. (The bill is still pending.)

You should supplement your assessment of potential major air and water pollutant release with an estimate of possible toxic pollutant release. The *Industrial Pollution Control Handbook*, mentioned above, is a good source of information about potential plant toxic air and water emissions. Most major industries' toxic air emissions are also discussed quite completely in a mammoth 1973 EPA study entitled *Air Pollution Engineering Manual* (GPO #4.9:40/2, $14.50). An excellent reference on various industries' toxic water discharges is the *Water Quality Criteria Databook, Vol. I: Organic Chemical Pollution* and *Vol. II: Inorganic Chemical Pollution* (EPA #180-10DPV12/70). Although this book is unfortunately out of print at the Government Printing Office, you may be able to use it at an environmental or a pollution control agency library. An Army Corps of Engineers booklet entitled *Permits for Work and Structures in, and for Discharges or Deposits into Navigable Waters* (U.S. Army Corps of Engineers, Washington, D.C., free) lists all significant types of water pollutants, including toxic contaminants, expected to be released from production operations in 51 industries, although no attempt is made to quantify the possible amounts.

Potential Solid Wastes

In addition to examining potential air and water pollution problems at the factory, you should consider the potential for generation of solid wastes. Most factories create large amounts of solid waste, including old machinery, leftover scraps of raw materials or products, "slag" or "tailings" which are by-products of production operations, ash and dust removed from exhaust gases by air pollution control equipment, and sludge captured by water treatment facilities. For example, about 800 pounds of inert grayish rock-like slag are produced for every ton of pig iron made in a blast furnace. Over the last century, about 90 million tons of slag have accumulated into one huge mountain near the steel mills of Pittsburgh, Pennsylvania.

Your research into industry technology and pollution load may have already suggested where to expect solid waste problems at the factory you are studying. The amount of solid waste produced can vary greatly with plant housekeeping procedures, recycling efforts, and the type and efficiency of existing air and water pollution control equipment. A power plant sulfur dioxide control system, known as a "limestone scrubber," produces large quantities of a virtually useless liquid-solid mixture, for example; however, an alternative control system may result in no such disposal problem, since the by-product is sulfur, which can often be sold and recycled.

Potential Off-site Impact

To understand more about a factory's "total" potential environmental impact, it is useful to consider briefly the ways its operations may affect the environment at locations other than the plant site. The power used by aluminum smelters in the Pacific Northwest, for example, is produced by hydro-electric dams which require the flooding of large areas. Some of the iron ore used at certain iron and steel mills in this country is processed at a plant which dumps 67,000 tons of tailings into Lake Superior each day.

While the company operating a factory is generally not directly responsible for off-site environmental impact, your awareness of such problems will prove valuable in considering the "trade-offs" involved in reducing pollution at the plant itself, as discussed under "Developing Pollution Control Proposals" in Chapter 4. Certain types of less-polluting steel-making furnaces use a greater proportion of scrap, for example, cutting down on ore demand.

If a company were considering replacing furnaces, this might be the most desirable option in terms of total environmental impact. On the other hand, it would not make sense in terms of total impact for a plant to try to cut down on particulate pollution by switching from burning coal to buying electric power if the electric utility also had to burn coal to produce the power.

At this point, you may simply want to make a list of the kind and amount of raw materials, including fuels and electricity, the factory you are analyzing uses, and match them with the plants (if you know them), or the industries, which produce them. While referring to some of the general information sources outlined above, you can then make note of any major potential air and water pollution, solid waste or land use problems of which you become aware in these industries.

EFFECTS OF POLLUTION

Each of the air and water pollutants discharged from a factory may have certain effects on the environment. It is very important to know something about these effects and the levels at which they occur, since this will give you a framework within which to evaluate plant emissions, and weigh control priorities.

Some pollutants, like arsenic, may persist for a long time; others, like cyanide, decompose relatively rapidly. A large quantity of a pollutant like iron may have no effect on humans, while it would kill all plant or fish life contacting it. Certain pollutants, like mercury and DDT, tend to concentrate in the food chain, and thus become more dangerous to life years after their release than they were initially. Some pollutants have one effect when released in large amounts for a short period, and another when released continuously in small amounts. Sulfur dioxide, for example, causes choking, tearing, and sneezing in concentrations of over 20 parts per million, and can contribute to lung disease in concentrations of as low as .2 parts per million over a long period of time.

Five major air pollutants, particulates, sulfur dioxide, hydrocarbons, nitrogen oxides, and carbon monoxide, are produced, in greater or lesser amounts, by industrial processes and fuel burning. **Particulate** pollution is a catch-all term for any solid bits of soot or dust or liquid droplets floating in the air. **Sulfur dioxide** is a pungent, irritating colorless gas which tends to react in the air with oxygen and moisture to form acid droplets and other "sulfates."

Some forms of **nitrogen oxide** gases are invisible; others have a yellow-brown color. Nitrogen oxides easily react in the presence of sunlight with hydrocarbons, to form photochemical

Polluted area on Gulf Coast, 100 miles south of Houston, Texas.

smog. **Hydrocarbons** are all substances containing carbon and hydrogen; gasoline and natural gas are common members of this family. Hydrocarbons may be gases, liquids or solids; however, it is generally only the gases which pollute the air.

Carbon monoxide is an odorless, invisible gas which is dangerous only to red-blooded animals, as it replaces oxygen in the blood.

Water pollution is frequently described in terms of a few aggregate measures such as **dissolved** and **suspended** solids, the amount of solid material in the water; **pH,** a partial measure of how acid or alkaline water is, generally expressed on a scale of 1 (most acid) through 7 (neutral) to 14 (most alkaline); and **biochemical oxygen demand,** a measure of the amount of oxygen bacteria will use up, in decomposing organic wastes.

Water pollution is also measured in terms of the exact amount present of specific pollutants, such as oil, phenol, cyanide, and any of a large number of metals. Some of these are relatively harmless but others can disrupt fish reproductive cycles, kill aquatic plants and animals, impart a foul taste to water, or render it totally unfit for drinking due to toxic or carcinogenic properties, depending on the pollutant and its concentration.

Your investigation into the "Pollution Potential" of the factory you are concerned about should have indicated which major and minor (but toxic) pollutants it may discharge.

Sources of Information

There are a great many sources of information about the environmental and health effects of pollution, ranging from simple, illustrated pamphlets published by the EPA to sophisticated scientific reports published in academic journals. *Air Pollution Primer*, available from the American Lung Association (N.Y. free), *Cost of Air Pollution Damage* published by the EPA (GPO#EP4 9:85, Washington, D.C., $.70), *A Primer on Waste Water Treatment,* also by EPA (GPO #0-419-407, Washington, D.C., $.55) and *Cleaning Our Environment—The Chemical Basis of Action* by the American Chemical Society (Washington, D.C., $2.75), are all good places to begin reading about pollution's effects.

Two more detailed, comprehensive and authoritative references on the health effects of pollutants are Volume 1 of Arthur Stern's *Fundamentals of Air Pollution* (Academic Press, New York, $14.50), and a book called *Water Quality Criteria* (GPO #EP 1.23:73-033, $12.80), published by the Federal Water Quality Administration in 1968, and commonly known as the "green book." (EPA was scheduled to release a "blue book" based on a National Academy of Sciences survey sometime in 1975, which would replace this reference.) These references should be available in libraries.

An excellent source of additional data on how air pollutants cause various environmental problems is the federal government. Prior to the advent of the EPA, the U. S. Dept. of Health, Education and Welfare, in 1969 and 1970, published studies which discuss the health effects of the five air pollutants, called "criteria pollutants," mentioned above. These highly factual, but readable reports are titled *Air Quality Criteria For (name of pollutant)* (order # varies, $1.50—1.75, see list).

AIR QUALITY CRITERIA REPORTS

Title	Order No.	Price
Air Quality Criteria for Particulate Matter	NTIS # PB–190–251	obtain from NTIS
Air Quality Criteria for Sulfur Oxides	NTIS # PB–190–252	obtain from NTIS
Air Quality Criteria for Hydrocarbons	NTIS # PB–190–489	obtain from NTIS
Air Quality Criteria for Carbon Monoxide	GPO # HE–20.1309:62	$1.50
Air Quality Criteria for Nitrogen Oxides	GPO # EP–4.9:84	$1.50

Based on available research into the health effects of the "criteria" pollutants the EPA has established two sets of limits, called "primary and secondary ambient air standards." Primary stand-

Table 1

NATIONAL AMBIENT AIR QUALITY STANDARDS

Pollutant	Primary	Secondary
PARTICULATE MATTER		
Annual geometric mean	75	60
Maximum 24-hour concentration*	260	150
SULFUR OXIDES		
Annual arithmetic mean	80 (.03ppm)	
Maximum 24 hour concentration*	365 (.14ppm)	
CARBON MONOXIDE		
Maximum 8-hour concentration*	10 (9ppm)	
Maximum 1-hour concentration*	40 (35ppm)	same as primary
PHOTOCHEMICAL OXIDANTS		
Maximum 1-hour concentration*	160 (.08ppm)	same as primary
HYDROCARBONS		
Maximum 3-hour (6-9 am) concentration*	160 (.24ppm)	same as primary
NITROGEN OXIDES		
Annual arithmetic mean	100 (.05ppm)	same as primary

[All measurements are expressed in micrograms per cubic meter (ug/m^3) except those for carbon monoxide, which are expressed in milligrams per cubic meter (mg/m^3). Equivalent measurements in parts per million (ppm) are given for the gaseous pollutants.]

* Not to be exceeded more than once a year.

ards specify the maximum concentration in the environment which will not be injurious to human health. Secondary standards, which may be more stringent, specify the levels of pollution which cannot be exceeded to protect public welfare and prevent damage to property and vegetation. Both sets of standards are called "ambient" because they specify the concentration of a pollutant allowable in the air, rather than the amount of pollution allowed to be emitted from a smokestack. (The latter are called "emission limitation" or "stack gas" standards.) The primary and secondary standards for the criteria pollutants are listed in Table 1.

In addition to the criteria pollutants, you may also have found that certain toxic air and water pollutants may be emitted from the factory you are studying. For 30 of the most significant toxic air pollutants (see list), the National Air Pollution Control Agency in 1969 prepared reports describing environmental effects, where the pollutants are likely to occur, and some of the ways to control them. Each study is entitled *Air Pollution Aspects of (name)* and is available from the National Technical Information Service (NTIS # varies, $6.00 each).

The Office of Toxic Substances at the EPA so far has promulgated emission limits for only three toxic air pollutants: asbestos, mercury and beryllium, and has issued no final standards for any toxic water pollutants. It did, however, issue a list of 123 toxic water pollutants in September, 1974, and has proposed standards for nine, in December, 1973. The promulgated air and proposed water standards appear in Table 2.

The National Institute for Occupational Safety and Health (NIOSH) has also conducted research since its inception in 1970, into the health effects of industrial air pollutants. As of October, 1974, it had published *Toxic Substance Criteria Documents* for about 20 workplace pollutants (see list, GPO # varies, $.95-$2.10), each summarizing research to date into the substance's effects.

Eventually the Federal Occupational Safety and Health Administration (OSHA) should promulgate workplace standards for each of these substances, although as of early 1975 it had set limits only for asbestos, vinyl chloride, and a group of carcinogenic

**TOXIC POLLUTANTS FOR WHICH
"AIR POLLUTION ASPECTS OF (name of pollutant)"
REPORTS HAVE BEEN PREPARED**

Pollutant	NTIS Order No.
Aeroallergens (pollens)	PB 188–076
Aldehydes (including acrolein and formaldehyde)	PB 188–081
Ammonia	PB 188–082
Arsenic and its compounds	PB 188–071
Asbestos	PB 188–080
Barium and its compounds	PB 188–083
Beryllium and its compounds	PB 188–078
Biological Aerosols (microorganisms)	PB 188–084
Boron and its compounds	PB 188–085
Cadmium and its compounds	PB 188–086
Chlorine gas	PB 188–087
Chromium and its compounds (includes chromic acid)	PB 188–075
Ethylene	PB 188–069
Hydrochloric acid	PB 188–067
Hydrogen sulfide	PB 188–068
Iron and its compounds	PB 188–088
Manganese and its compounds	PB 188–079
Mercury and its compounds	PB 188–074
Nickel and its compounds	PB 188–070
Odorous compounds	PB 188–089
Organic carcinogens	PB 188–090
Pesticides	PB 188–091
Phosphorus and its compounds	PB 188–073
Radioactive substances	PB 188–092
Selenium and its compounds	PB 188–077
Vanadium and its compounds	PB 188–093
Zinc and its compounds	PB 188–072

Reports available from National Technical Information Service, price $6.00. When ordering, refer to title as "Preliminary Air Pollution Survey of (name of pollutant)."

Table 2
NATIONAL TOXIC SUBSTANCE STANDARDS

AIR POLLUTANTS—PROMULGATED

(see Federal Register of 4/6/73 for full description of requirements)

Pollutant	Limit
asbestos	no visible emissions
mercury	2,300 grams/day from a stationary source
beryllium	10 grams/day OR level such that ambient concentration in vicinity of source does not exceed 0.01 ug/m^3 averaged over a 30-day period

WATER POLLUTANTS—PROPOSED

(see Federal Register of 12/27/73 for full description of limits; standards may be lower, depending on flow of discharger and receiving water body)

Pollutant	Maximum Discharge Concentration in Effluent [in micrograms per liter (ug/l)]* Fresh Water	Salt Water
Aldrin-Dieldrin	0.5	5.5
Benzidene	1.8	1.8
Cadmium	40.0	320.0
Cyanide	100.0	100.0
DDT, DDE, DDD	0.2	0.6
Endrin	0.2	0.6
Mercury	20.0	100.0
PCB's	280.0	10.0
Toxaphene	1.0	1.0

* equivalent to parts per billion (ppb)

NIOSH TOXIC SUBSTANCE "CRITERIA DOCUMENTS"

Pollutant	GPO Order No.	Price
Ammonia	1733-00036	$1.55
Arsenic (inorganic)	1733-00030	1.50
Asbestos	1733-00009	2.10
Benzene	1733-00038	2.10
Beryllium	1733-00011	2.10
Carbon Monoxide	1733-00006	2.00
Chloroform	1733-00045	1.95
Chromic Acid	1733-00020	1.10
Coke Oven Emissions	1733-00014	.95
Cotton Dust	1733-00044	2.35
Lead (inorganic)	1733-00013	1.25
Mercury (inorganic)	1733-00022	1.50
Sulfur Dioxide	1733-00029	1.55
Sulfuric Acid	1733-00034	1.40
Toluene	1733-00019	1.25
Toluene Diisocyanate	1733-00021	1.25
Trichloroethylene	1733-00023	1.30
Ultraviolet Radiation	1733-00012	1.25

organic chemicals. In the meantime, it is enforcing exposure limits established by a non-governmental group called the American Conference of Governmental and Industrial Hygienists. These limits, indicating for hundreds of toxic air pollutants the concentrations known to be dangerous to workers, appear in a booklet called *Threshold Limit Values for Chemical Substances in Workroom Air* (ACGIH, Cincinnati, Ohio, $.75).

All the references listed above are summaries, simple to detailed, of the results of research into the environmental effects of pollutants. Eventually, you may want to dig up some original research reports yourself, once you have done investigation of actual factory emissions and isolated which pollutants present problems. Some of the most useful articles about environmental studies are published in science magazines directed to the layperson, such as *Scientific American* (New York, $15), *Science* (American Association for the Advancement of Science, Washington, D.C., $50), *Environment* (Bridgeton, Mo., $10), and *Environmental Science and Technology* (American Chemical Society, Washington, D.C., $9). The trade journals in the industry also occasionally publish articles about environmental effects of pollution. All of these magazines are indexed in the *Readers' Guide* or *Funk and Scott*.

It is also possible to go further—directly to the papers published by the scientists performing research. Several indexes can be used to track these reports down. The Air Pollution Technical Information Center (APTIC) of the EPA publishes semi-annually a basic index called *Air Pollution Technical Publications of the U.S. Environmental Protection Agency*. This index, which gives the name, GPO or NTIS number, and cost of each document is available free of charge from APTIC, Research Triangle Park, North Carolina. EPA's Office of Research and Development periodically publishes another index entitled *Bibliography of R&D Research Reports*. Subscriptions to this index can be obtained free from that office in Washington, D.C. The Office of Water Resources Research (OWRR) of the U.S. Dept. of the Interior publishes a quarterly index called *Research Reports* listing OWCC supported publications and a semi-monthly index called *Selected Water Resources Abstracts*. The latter source is particularly useful because it contains brief summaries of all indexed reports. Subscriptions to both indexes can be arranged through the Water Resources Scientific Information Center of the Department of Interior in Washington, D.C.

Articles about solid waste are indexed in EPA booklets called *Available Information Materials* (Solid Waste Information Materials Control Agency, Cincinnati, Ohio, free).

Reports published by the National Institute for Occupational Safety and Health are indexed in its *Current Publications*. This is updated periodically, and available free, from NIOSH's Office of Technical Publications (Post Office Building, Room 530, Cincinnati, Ohio), if you write and ask to be placed on the mailing list.

Besides reading technical papers and their summaries, you may at some point wish to consult or correspond with some of the experts in the field. Most scientific articles list the author's affiliation through which a line of communication can be established. In January 1974, the EPA published a list of its key research personnel, called *National Environmental Research Center Program Directory*, (Office of Program Management, EPA, Washington, D.C., free), which provides the names, addresses, phone numbers, job titles, and fields of research for over a thousand pollution specialists. The Conservation Foundation has also published an extensive list of scientists (at least a dozen from each state), who have expressly stated their willingness to answer questions from the public about technical aspects of environmental issues. The list, with addresses and phone numbers, is contained in its *Water Quality Training Institute Information Kit* (Washington, D.C., $10, $6 for non-profit groups). (This kit, which fills a very large looseleaf binder, contains much valuable information on the NPDES water program as well.)

POLLUTION CONTROLS

There are a number of ways to minimize or prevent the formation or release of almost all air and water pollutants. If your aim is to understand plant problems and to assess how a company may reduce its pollution, you must be familiar with available pollution control methods.

These methods fall into six broad categories. The company can:

    (1) install devices specifically designed to reduce or prevent pollution

    (2) improve in-plant maintenance and housekeeping procedures

(3) change raw materials or fuels to ones which create less pollution

(4) conserve energy

(5) alter the production process so that it produces the same product while generating less pollution, or

(6) shut down a highly-polluting operation altogether.

These options can be employed separately or, frequently, in combination, to control a factory's air, water and solid waste problems.

Air Pollution Control Systems

The first option—that of installing pollution control devices—is the one that generally comes to mind when speaking of pollution control. There are four major types of air pollution control systems which can be added to most industrial operations to clean "stack" or "exhaust" gases.

Electrostatic precipitators have served as the workhorses of particulate control for over fifty years. Improved over time, they can now remove up to 99.5 percent of soot and dust emissions at some factories. A precipitator consists of a boxlike structure, often several stories high, containing electrically charged plates that create an electric field. As the gas stream passes through it before going up the stack, soot particles are attracted to the plates. From time to time, the plates are "rapped" and the clinging particulate matter falls to the bottom of the precipitator.

Baghouse filters can be even more efficient at controlling particulates, trapping up to 99.9 percent of soot and ash at some plants, including the tiny submicron-size particles which may be most damaging to the environment and to human health. "Baghouses" are large structures containing woven bags. Before going up the smokestack, the flue gas is forced through the holes between the bag fibers. Soot and ash bits cling to the fabric, much as in a vacuum cleaner. Periodically, the collected particles are removed by shaking out the bags.

Wet scrubbers are the third standard type of particulate control device. They can be designed not only to remove 99.5 percent of soot particulates but also 80 to 95 percent of sulfur dioxide as

Electrostatic precipitator (box-like structure) atop power boilers at paper mill in Texarkana, Texas.

well. They operate by sending the flue gas stream through a liquid spray. The particles and polluting gases in the flue are absorbed by the liquid, which is then flushed away to a water treatment unit. Although several large coal-burning electric utilities have strongly challenged the effectiveness and practicality of wet scrubbers to remove sulfur dioxide at power plants, claiming that the scrubbers do not prove consistently operable and that the "sludge" wastes from the system may create a massive disposal problem, such systems are achieving increasingly widespread acceptance. As of December, 1974, 22 coal-burning power plants had sulfur dioxide scrubbers already in operation and 71 had them under construction.

Electrostatic precipitators, baghouse filters and wet scrubbers can all cost several million dollars each. A fourth type of air pollution control device long used by industry, known as a "mechanical" or "cyclone" collector, costs far less. However, **mechanical collectors** are not only the simplest but least effective of these systems. Collecting particles solely by mechanical means, such as passing the flue gas through a funnel-shaped chamber where soot and dust will settle out, the best-designed systems capture little more than 95 percent of large particles, and a far lower percentage of smaller ones. While mechanical collectors are still frequently used along with electrostatic precipitators in a "precleaning" function, even this use is more and more considered obsolete.

Finally, in an effort to control the concentrations of air pollutants at ground level, some companies have built very **tall smokestacks** —up to a thousand feet high. This does not constitute a pollution control method, however, but is rather a method of pollution dispersion. Tall stacks, along with "intermittent controls," a term used to describe the practice of cutting back production when weather conditions threaten to raise local pollution levels above ambient air standards, have been found by EPA and the courts to be unacceptable permanent methods of complying with the Clean Air Act. However, the EPA has approved their use on an interim basis for certain coal-burning power plants and copper smelters, while adequate control technologies are being developed.

Beyond addition of devices to smokestacks, there are some **modifications of fuels or processes** used in a plant that can greatly help reduce air pollution levels. Gaseous pollutants in particular are commonly controlled by such methods.

The EPA estimates that over 75 percent of all U.S. sulfur dioxide pollution results from the burning of coal and oil. By limiting these fuels, which may consist of as much as 4 percent sulfur, to types with a "low" sulfur content (generally defined as containing less than one percent sulfur, although some seriously polluted localities restrict the sulfur content of fuels to a few tenths of a percent), the amount of sulfur dioxide generated during combustion can be reduced. However, low-sulfur fuels can be more expensive and, in some cases, are now unavailable because of high demand.

Engineers have also found that emissions of some pollutants, such as nitrogen oxides, can be controlled by design changes in production equipment which either prevent the pollutants' formation or cause them to burn or decompose into harmless by-products before they are released into the air. California's two major utilities, Southern California Edison and Pacific Gas and Electric Company, reduced the nitrogen oxide emissions from their gas-burning power plants by 70 percent through changes in boiler operation.

Water Pollution Control Systems

Like air pollutants, water pollutants can be prevented from contaminating the environment either by utilization of "cleaning" devices which remove the pollutants from process water, or by modifications of the process, which prevent their ever entering the water. Three types of treatment systems are commonly employed to purify the water a factory has used, prior to returning it to a river, lake, ocean or other waterway, and two types of treatment are used to remove waste heat.

Primary treatment systems use strictly physical processes to remove up to 90 percent of the heavy settleable solid material in plant effluent. These treatment facilities—all essentially large tanks or ponds—are known as settling basins and clarifiers. In most of them, the plant effluent sits while the largest, heaviest waste material sinks to the bottom and the clear water slowly overflows the top. Buoyant wastes such as oil will float to the surface where they may be mechanically skimmed off.

Secondary treatment systems use biological processes to remove up to 95 percent of the organic material in the effluent. This equipment is more varied and expensive than that for primary treatment.

The most commonly-used secondary systems consist of "holding ponds," large enough to retain a plant's effluent until bacteria can decompose the wastes. Wastes are held for days, weeks, or even months—in such systems. In some ponds, however, the process of biological breakdown is speeded up by forcing oxygen into the effluent with aerators. Since the waste-consuming bacteria use up oxygen in the decomposition process, this allows them to grow and function much more rapidly.

Two other more elaborate, quick, and compact types of secondary treatment systems, known as "trickling filters" and "activated sludge," are used by many municipal sewage and industrial plants. A trickling filter consists simply of a bed of stones, three to ten feet thick, on which bacteria live and multiply. Polluted water is sprayed over the stones, and as it trickles down through the rocks, the bacteria decompose the biological wastes. In an activated sludge system, the "sludge"—a mixture rich in bacteria—is whipped into the waste water by giant aerators. The bacteria quickly attack and break down the wastes so that within a few hours, instead of a few days or weeks, the effluent can be discharged. Advanced secondary treatment systems have also been developed for unusual purposes. Bethlehem Steel's "bacteria cafeteria" at its Bethlehem, Pa. mill, for example, utilizes a specially-developed strain of bacteria to devour and decompose poisonous cyanide wastes.

Tertiary treatment is a catch-all phrase for anything beyond secondary, involving use of chemical methods and advanced physical techniques. The process depends on the nature of the wastes. In municipal sewage, tertiary treatment removes nitrates and phosphates; in pulp and paper industry effluent, it removes coloring matter and organic compounds; in steel industry effluent, it removes heavy metals.

Aeration lagoon (bottom), holding pond (top left) and clarifier (top right) at pulp mill. White pattern in lagoon is made by air bubbled into water.

While the primary, secondary and tertiary water treatment processes can eliminate most waste materials suspended or dissolved in effluent, they do not have any effect on the separate problem of "thermal pollution," created when water heated during factory use is discharged into cooler rivers or other waterways. Thermal pollution control, the cooling off of heated water prior to discharge, is accomplished by use of cooling ponds or cooling towers.

Cooling ponds, the simplest types of thermal control systems, are man-made lakes in which the heated water from a plant slowly circulates until it is cooled—primarily by evaporation. For relatively large—1,000 megawatt—power plants, however, which have serious thermal pollution problems, ponds generally would need a surface area of 2 to 3 square miles. This tremendous land requirement clearly limits pond use, although use of spray devices, which aerate the water by shooting it over the surface, allows a reduction in surface area by as much as a factor of 20.

Cooling towers, more elaborate and expensive thermal control systems but ones requiring much less land area, function by providing a structure in which heated water is cooled by being brought into contact with moving air. In "wet towers," the hot water falls in a thin film over a series of baffles and is cooled primarily by evaporation. In "dry towers," designed mainly for use in water-short areas, the hot water is circulated inside a large series of closed pipes; heat is transferred from the water to the air by convection and conduction rather than evaporation.

By using a combination of pollution control techniques, some companies have succeeded in cleaning process water well enough to recycle it through a plant again and again. Several large steel mills, including Republic's South Chicago Works, which formerly discharged thousands of pounds of pollution a day, are now moving toward 100 percent **recycling of wastewater.** The process of "closing the loop" completely prevents any water pollution from leaving a factory.

The cost of water pollution treatment systems can vary with their size and complexity from a few thousand to hundreds of thousands of dollars.

Cooling tower for nuclear power plant on Columbia River near Prescott, Washington.

Solid Waste Controls

The extraction of pollutants from stack gases and waste water often adds to the accumulation of large amounts of solid and liquid wastes at a plant. Devices and systems have been developed for disposing of these and other wastes. Industrial by-products can sometimes be sold to industries for **recycling or reprocessing.** Slag from steel mills which had been accumulating for a century is today being used to make concrete and mineral wool insulation. The sulfuric acid produced by sulfur dioxide control systems in copper smelters is often sold, or reused in ore-processing operations.

While recycling is by far the most desirable method of **solid waste disposal,** much waste unfortunately has little value, particularly the ashes and sludges captured by air and water pollution control systems. These should be disposed of in a place where they will cause the least possible environmental harm, particularly if they are contaminated with toxic substances. Most commonly, solid waste materials are dumped in a landfill or sanitary landfill (where each day's waste is covered with dirt), in a deep well in the ground, in the ocean or a lake, or in an abandoned quarry or mine. They may also be burned in an incinerator.

Potential dangers accompany all these methods: groundwater at a landfill site may become contaminated, deep wells can leak, wastes dumped in waterbodies can injure aquatic life and come back up on beaches, and incineration can cause air pollution. The least harmful method for your locality has to depend on local conditions.

Alterations in Operations to Control Pollution

Besides adding pollution control devices, changing materials or fuels, recycling, and disposing of wastes, a company can reduce plant pollution in several other ways. Simply **improving house-keeping and maintenance** operations can often make a significant difference. A major source of air pollution at an oil refinery, for example, is leaks from the hundreds of miles of pipes running through it. If maintenance crews spot and replace worn pipes and weakened joints before leaks develop, air pollution from this source can be prevented. Proper housekeeping practices can be the most important method of water pollution control in the food processing industry. At apple processing plants, for example, heavy water pollution discharges can be prevented simply by sweeping up, rather than hosing down or sluicing away wastes.

Energy conservation measures can also lessen plant pollution by reducing the amount of fuel burned, and thus the emissions that would result from this burning. Industrial energy conservation measures, while different at every plant, can include reductions in lighting and heating, use of more efficient motors, recovery and reuse of waste heat, improved insulation of walls and pipes, conversion of continuously operating back-up systems to automatic standby control, and shutting down of machinery when not in actual use.

An Upjohn plant in Kalamazoo, Mich. achieved a 16 percent reduction in the amount of steam it uses—produced by burning coal—simply by cutting warehouse temperatures to 65 degrees and keeping other areas at 68. In another case, an Alcoa aluminum plant in Warrick, Ind. was able to save more than 60 million kilowatt hours of electricity a year, an amount equivalent to the power use of about 7,500 homes, by switching from use of a "mercury arc" to "silicon diodes" in its production process.

A much more complicated and expensive approach to reducing pollution is to install **a new production process** that makes the same product, but in a less-polluting way. For example, there are two generally accepted technologies for producing chlorine: the mercury cell and the diaphragm cell. Until 1970, most manufacturers at the nation's 200 chlor-alkali plants used the mercury cell, which had the potential to release six pounds of mercury per ton of chlorine produced into the factory's waste water. However, in the last five years, in response to generally heightened environmental concern as well as to fears regarding mercury contamination of fish, most U. S. manufacturers of chlorine in the country have switched to the diaphragm cell process, which creates no mercury pollution.

If no other alternatives seem feasible and environmental problems seem overwhelming or too costly to correct, a company might choose to **stop producing a product,** or to **close down a particular plant**. Dow Chemical Company recently claimed that it has instituted a "product stewardship" program which entails reviewing all of its production processes and products in terms of environmental impact. Although citing no examples, the company says that it will stop manufacturing some products if environmental considerations are found to outweigh product profitability. (Dow also says that it reviews the way industrial buyers use certain of its products in their manufacturing processes, according to environmental criteria.)

Sources of Information

With over 1500 companies now manufacturing pollution control equipment, there is certainly no shortage of pollution control options available to minimize factory problems. However, those applicable for a particular plant depend on the factory production processes, the size, age and complexity of its operations, and pollution control efforts to date, among other factors. To assess which would be best for the factory you are concerned with, you must develop a basic grasp of control techniques widely used in the industry, and of those which might suit the plant. You can begin learning about the operation of different pollution control systems by reading two good booklets: the EPA's *A Primer on Waste Water Treatment*, mentioned above, (GPO # 0-419-407, $.55), and *Controlling Air Pollution*, by the American Lung Association, (New York, free).

More detailed discussions of the technologies and their costs appear in Stern's *Fundamentals of Air Pollution*, the EPA's *Air Pollution Engineering Manual*, and Lund's *Industrial Pollution Control Handbook*, all mentioned above. These books provide clear explanations, and at least one should be available in a general or engineering library.

To understand next the systems applicable to a particular industry, you may want to obtain and digest materials available from the EPA. Five HEW reports, called *Control Techniques For (pollutant)* (order # varies; $.70-$2.00, see list), include analyses of all available control techniques for each of the five

AIR POLLUTION
CONTROL TECHNIQUES REPORTS

Title	GPO Order No.	Price
Control Techniques for Particulate Air Pollutants	FS 2.93/3:51	$1.75
Control Techniques for Sulfur Oxide Air Pollutants	FS 2.93/3:52	$2.00
Control Techniques for Carbon Monoxide Emissions from Stationary Sources	HE 20.1309:65	$ .70
Control Techniques for Hydrocarbons and Organic Solvent Emissions from Stationary Sources	HE 20.1309:68	$1.00
Control Techniques for Nitrogen Oxide Emissions from Stationary Sources	HE 20.1309:67	$1.00

"criteria" air pollutants. You should read those reports discussing the pollutants which are problems in the industry you are studying.

In the process now underway of developing air pollution standards for new plants in certain industries, EPA has prepared *Background Information* reports which extensively discuss industry air control technology options. Eighteen of these have been released to date (NTIS # varies, $3.00-$4.50, see list). Discussions of water pollution control options appear in the *Development Documents*, prepared as part of EPA's water pollution standard setting program, mentioned above. If *Background Information* and *Development Documents* have been drawn up for the industry you are analyzing, they may be the most valuable basic source available on the industry's air and water control technologies.

Some supplementary materials describing industry control alternatives may also be available from industry trade associations or from the trade associations of pollution control equipment manu-

facturers: the Industrial Gas Cleaning Institute (Stamford,
Conn.), the Water and Wastewater Equipment Manufacturers
Association (Newark, N.J.) and the Environmental Equipment
Institute (Little Neck, N.Y.).

To be most up-to-date on pollution control equipment, you may
want to check through recent issues of magazines in the field for
articles which apply to the industry. Two of the best for the lay-

---

AIR POLLUTION
BACKGROUND INFORMATION
FOR PROPOSED NEW SOURCE PERFORMANCE STANDARDS DOCUMENTS

Industry	NTIS Order No.	Price
Steam Electric Generators	PB 202-459	$3.00
Portland Cement Plants		
Incinerators		
Nitric Acid Plants		
Sulfuric Acid Plants		
Asphalt Concrete Plants	PB 221-736	$4.50
Petroleum Refineries		
Storage Vessels (Petroleum)		
Secondary Lead Smelters and Refineries		
Brass or Bronze Ingots Plants		
Iron and Steel Plants		
Sewage Treatment Plants		
Coal Preparation Plants	free on request from the	
Primary Copper, Lead and Zinc Smelters	U.S. Environmental	
Primary Aluminum Plants	Protection Agency	
Steel Plants: Electric Arc Furnaces		
Ferroalloy Production Facilities		
Phosphate Fertilizer Industry		

Solid wastes: slag heap at coal processing facility.

person are the monthly *Environmental Science and Technology,* mentioned above, and the weekly *Air/Water Pollution Report* (Silver Spring, Md., $145 a year). The *Journal of the Air Pollution Control Association* (Pittsburgh, Pa., $25 a year) and the *Journal of the Water Pollution Control Federation* (Washington, D.C., $22 a year) also contain excellent, though more technical and detailed articles. All can be found in most engineering libraries. The numerous indexes of business and trade magazines mentioned above can also be used to locate articles about industry pollution control systems in these and other technical magazines.

Not all industry control systems will be appropriate for the factory with which you are concerned. Before you can sort out the most workable alternatives, you will need to gather certain specific data, described in Chapter 3, and discuss control possibilities with a variety of experts in the field. However, you can begin learning now about which systems might work best, from companies which manufacture and design them. (You can then return to these sources later for more specific discussions of control alternatives.) Most pollution control equipment manufacturers have published booklets describing their products. Someone in the business can sometimes provide a better understanding of plant options in an informal interview. The names, addresses and product lines of pollution control equipment manufacturers are indexed in an annual special issue of the bi-weekly magazine, *Chemical Engineering.* (Single copies of this issue, called the "Environmental Engineering Deskbook," can be ordered from the magazine in New York City for $2.50.)

Since the advent of stricter environmental regulations, the field of environmental consulting has been booming. Many consulting firms now advise extensively or solely on environmental problems and solutions. Among the major firms are Arthur D. Little in Cambridge, Mass., Battelle Memorial Institute in Columbus, Ohio, Datagraphics in Pittsburgh, Pa., and the Stanford Research Institute in Palo Alto, Calif. Xerox Corporation published a book called *Environment USA* (Bowker Order Dept., Ann Arbor, Mich., $15.95), a comprehensive directory of people and organizations active in environmental matters, which lists the names of 850 consulting firms and individuals specializing in environmental problems.

Although consultants generally offer their expertise to corporations and the government for a fee, you might find a sympathetic specialist willing to discuss pollution control technology with you. You might also find that some of the results of research projects done by these firms have been published and are obtainable on simple written request, although others may be priced anywhere up to $1,000.

ECONOMICS OF POLLUTION CONTROL

The installation of pollution controls is a costly affair. Many single pieces of control equipment cost over a million dollars. The President's Council on Environmental Quality has estimated that American industry will spend a total of $107 billion between 1973 and 1982 on air and water pollution abatement.

Some companies can readily bear the costs of cleaning up their factories' effluents. Others, however, may find pollution control costs creating an additional financial burden that results in a factory's becoming unprofitable. To understand the feasibility of upgrading pollution controls at a particular factory, it is important to be aware of the costs involved in various cleanup strategies, and the potential repercussions of making such expenditures on the factory, the company and on local and national economies.

You should begin your research in this area by obtaining an overview of the relative financial health of the company, and of the industry of which it is a part, since this may suggest the amount of money available for pollution control programs. While such inferences may be complicated by the fact that a subsidiary's position may not be accurately reflected in that of the parent company, and by the fact that economic conditions are constantly changing, it is nevertheless important to at least be aware of the company's basic profitability and financial status.

The most accessible sources of information about companies' and industries' economic ups and downs are the *Wall Street Journal*,

The New York Times, and *Business Week,* and business monthlies such as *Forbes* (New York, $12 a year) and *Fortune* (Chicago, $14 a year). More specific information about the economic performance of an industry relative to others can be found in the April issue of the *Monthly Economic Letter* (New York, free), published by First National City Bank. This newsletter provides data on the profit margins and rates of return for each of 41 industries over the past two years. Trade organizations generally also publish voluminous material about the economic performance of their industry, which they should be willing to make available.

Every May issue of *Fortune* lists basic data on sales, profits, assets, growth and earnings per share for the 500 largest corporations in the United States; every June issue does the same for the second 500 largest corporations. (A special annual report of data from the May and June issues is available from *Fortune* in Chicago for $3.75). *Business Week* publishes charts quarterly listing various economic indicators for major U.S. corporations.

Once you are somewhat familiar with the financial health of the industry and company you are analyzing, you can begin to consider the factors affecting the company's expenditures for pollution control. You may want to start by looking briefly at the impact of pollution and control expenditures on the nation as a whole, since this will both provide you with considerable information on current environmental spending trends and acquaint you with the methods and terms economists use in evaluating and predicting their economic impact. Chapter Three of *Environmental Quality 1973: The Fourth Annual Report of the Council on Environmental Quality* (GPO # 4111-00020, $4.30) entitled "Economics and Environmental Management," is a good introduction to both the costs of pollution to society and the costs of pollution control to industry and to the public. The Conference Board, a private, industry-supported economic research organization, compiled a useful analysis in 1972 of cost factors related to water pollution control in industry, called *Industry Expenditures for Water Pollution Abatement* (New York, $3.50 for members and education groups, $17.50 for others).

Having placed pollution control expenditures in a national context, you can proceed to focus on studies which analyze the current effects, and predict the future impact, of pollution control expenditures on a particular industry. Numerous such analyses have been completed. The Council on Environmental Quality (CEQ) in 1972 sponsored economic impact studies for eleven manufacturing industries.* The studies are summarized in *The Economic Impact of Pollution Control: A Summary of Recent Studies.* (GPO # 0-458-471, $2.50). If the industry you are interested in is one of those covered, you should obtain this summary report and, if you intend to study economic factors in depth, the original study as well.

The *Development Documents* and the *Background Information* reports mentioned above also contain chapters discussing the cost of meeting proposed water and air standards for the industries they cover. In addition, companion reports to the *Development Documents* discuss effects of such costs on the industry (see list).

Ultimately, your research probably will lead you to conclude that the factory you are analyzing has certain options for improving its pollution control, and that planning and implementing each of these entails certain costs. If you have the time, interest and some economic expertise, you may want to investigate how the company which operates it could finance such expenditures.

Companies rarely pay for new pollution controls all at once; instead, since such equipment has a long expected lifetime, they generally finance such expenditures over a period of years. During this period, the company may have to borrow money—often by issuing a tax-free pollution control revenue bond—or reduce profits to pay for improvements. It may also decide to increase prices or possibly lower stockholders' dividends. Exactly how much it will have to spend will depend on a number of factors: the basic cost of engineering, building and installing the system, known as "capital" costs; annual operating and maintenance costs; costs, such as interest on loans, which the company incurs in raising money to buy the system; and cost savings obtained through tax breaks.

---

* The automotive, baking, cement, electric power, produce canning and freezing, iron foundry, leather tanning, non-ferrous metals, petroleum refining, pulp and paper, and steel industries.

ECONOMIC ANALYSIS OF PROPOSED EFFLUENT GUIDELINES
(companion reports to EPA Development Documents)

Industry	EPA No.
Asbestos: Building Construction & Paper	EPA–230/2–74–001
Beet Sugar	EPA–230/2–74–002
Cane Sugar	EPA–230/2–74–003
Cement Manufacturing	EPA–230/2–74–004
Dairy Product Processing	EPA–230/2–74–005
Steam Electric Powerplants	EPA–230/2–74–006
Electroplating: Copper, Nickel, Chrome & Zinc	EPA–230/2–74–007
Feedlots	EPA–230/2–74–008
Ferroalloys: Smelting & Slag Processing	EPA–230/2–74–009
Basic Fertilizer Chemicals	EPA–230/2–74–010
Insulation Fiberglass	EPA–230/2–74–011
Citrus, Apple & Potato Processing	EPA–230/2–74–012
Flat Glass	EPA–230/2–74–013
Grain Processing	EPA–230/2–74–014
Inorganic Chemicals: Major Inorganic Products	EPA–230/2–74–015
Leather Tanning & Finishing	EPA–230/2–74–016
Red Meat Processing	EPA–230/2–74–017
Non-ferrous Metals Manufacturing: Bauxite Refining, Primary & Secondary Aluminum Smelting	EPA–230/2–74–018
Organic Chemicals: Major Organic Products	EPA–230/2–74–019
Petroleum Refining	EPA–230/2–74–020
Phosphorous Derived Chemicals	EPA–230/2–74–021
Plastics: Synthetic Resins	EPA–230/2–74–022
Builders Paper & Roof Felt	EPA–230/2–74–023
Pulp & Paper: Unbleached Kraft & Semi-Chemical Pulp	EPA–230/2–74–023
Rubber Processing: Tire & Synthetic	EPA–230/2–74–024
Seafood: Catfish, Crab, Shrimp & Tuna Processing	EPA–230/2–74–025
Soap & Detergent Manufacturing	EPA–230/2–74–026
Steel Making	EPA–230/2–74–027
Textile Mills	EPA–230/2–74–028
Timber Products: Plywood, Hardboard & Wood Processing	EPA–230/2–74–029

These reports are available from the National Technical Information Service. Use the EPA number to obtain the NTIS order number and price before ordering.

An explanation of how these factors can be integrated to determine the "total" cost of various company pollution control options and the possible effects on prices and dividends given different methods of financing, appears in a study done by the Council on Economic Priorities, a non-profit research group. The study, entitled *Environmental Steel* (Praeger Publishers, New York, $18.50) applies its "economic model" outlined in Chapter 7 and Appendix II, to potential pollution control expenditures for seven steel companies. It explains to the layperson the economic principles involved, and contains a glossary of important economic terms.

SOCIAL CONTEXT AND CORPORATE PLANNING

A company's ability to pay for pollution controls is an extremely important determinant of how, when or even whether it will clean up a factory; however, it is not the only determinant. Less tangible factors, such as community and corporate "attitudes" toward pollution control, and overall corporate goals, will also affect its decisions.

One very important factor affecting whether a company invests in pollution controls at a particular factory is the social context in which the plant exists. In the Mahoning River valley in Ohio, for example, steel industry employment has dropped by half since World War II. Citizens, fearing the loss of more jobs, have picketed water pollution control agency meetings with signs reading STEEL NOT EELS. As late as 1974, the state of Ohio had done almost nothing to regulate industry discharges in the valley, and the Republic Steel Corp. was operating several of the worst water polluters in the entire U. S. steel industry there.

On the other hand, in Chicago, where enforcement has been relatively vigilant and environmental groups nothing short of militant, Republic has upgraded a large, old steel mill to the point where it will soon be recycling nearly all the water it uses, and discharging virtually no water pollution at all.

It is often useful to obtain some basic information about a company's role as employer, taxpayer and in some cases politician, since this may help explain both its importance to the community and its potential leverage in community affairs. Try and determine the total size of your city's workforce, and the percentage of it that works at the factory you are interested in; also the total city tax roll and the percentage which the factory pays. Union officials at the plant should have information on the plant workforce size. The municipal tax department will have tax records. *Who's Who,* carried in most libraries, provides information on community and other outside activities of high-level corporate officials. Politicians and newspaper reporters may also prove to be good sources of information on the corporate role in community affairs.

The corporation's overall policies and "attitude" on pollution control are also very important in determining what it has done and will do at a particular factory. Republic Steel appears to have had a policy of only cleaning up where strong community and legal forces demanded it. The Owens-Illinois paper division, on the other hand, pursued a policy of installing the best available pollution controls at all its pulp mills long before this was even suggested by legal agencies. The American Electric Power Company has practiced advanced thermal and particulate pollution control at most units. However, it has publicly taken the position that devices to capture sulfur dioxide in stack gases will not work at its plants and has built none.

You should be alert for mention of such company-wide trends or policies in business periodicals and newspaper articles. Positions on environmental issues are also sometimes mentioned in annual reports, although many companies limit themselves to general statements about being "for" a cleaner environment, without specifying exactly what such policies entail.

A third important factor in determining a corporation's actions at a particular plant, is how the corporation fits the plant into its long-term strategy for markets and growth. The plant may be marginal and due to be phased out shortly, or it may be a chief profit-maker scheduled for expansion. The plant's importance to the company will ultimately determine whether the firm decides to invest heavily in pollution control, install enough controls to keep the plant operating until stricter regulations take effect or replacement capacity has been built, or shut the plant down immediately.

It is generally possible to tell something about where a factory fits into company plans by looking at the other plants operated by the company: their age, size, location, major products, and whether any are being expanded or phased down relative to the factory you are interested in. This information may appear in annual reports and industry directories. Pollution control and company officials may also be willing to discuss such trends.

KEEPING IN TOUCH

Developments in pollution control are not ancient history. They are happening now. Through all stages of your research, data collection, analysis, and action, it is essential to keep in touch with changes in the field.

One way to stay abreast of new developments is to check in periodically with contacts you have made in your work: corporate officials, pollution control experts, EPA and state pollution control personnel, equipment manufacturers. Another is to keep up your reading of current issues of the trade magazines you have found most helpful. It is also worthwhile to get on the mailing list for press releases from the EPA, the state pollution control agency, the company's public relations department, and the industry trade organization, if possible.

The U. S. EPA publishes a highly useful monthly newsletter called *EPA Citizens' Bulletin* (EPA Office of Public Affairs, Washington, D. C., free) to which you should subscribe. It discusses important developments, lists enforcement actions by the EPA regions to implement Federal pollution laws, and indicates how to obtain more details. In most cases, it will refer you to specific issues of the *Federal Register*, a daily compilation of all official actions taken by Federal agencies. (Copies of the

Federal Register are available from the Government Printing Office at $.75 apiece.)

If a state agency publishes a newsletter, you may want to subscribe to it as well.

3

Gathering Factory Data

There is only one way to know whether an industrial plant is well controlled or a serious polluter, and that is to know exactly how much, and what kind, of pollution it is producing. Fortunately, national environmental legislation enacted in the early 1970s has made it possible for virtually anyone to obtain from the state and Federal governments, detailed statistics on almost any factory's air and water emissions. In addition, other important data, such as the efficiency and cost of existing factory pollution control systems, and the factory's legal status, can usually be, and should be, obtained from the company operating the factory, and from government enforcement agencies.

This chapter explains how and where to gather the specific figures and facts, including air and water pollution data, background information, and legal data, necessary to evaluate a factory's environmental problems and how they can be alleviated.

AIR POLLUTION DATA

The Law: The Clean Air Act

A comprehensive effort to reduce or prevent degradation of the nation's air began with Congressional passage of the Clean Air Act Amendments on December 30, 1970. The Clean Air Act targeted May 31, 1975 as the date by which all air in the country should be safe for every human being to breathe. To attain this goal the law called for the Federal government to set air pollution standards, and for states to develop strategies for their enforcement. In addition, the Act called for the collection of detailed information about the amount of pollution being emitted from all sources.

The Amendments represented a radical change from previous pollution control strategies. Three previous attempts at meaningful legislation, in 1955, 1963, and 1967, had authorized funds for pollution research, and for development of state regulatory agencies. The 1970 Act granted the newly formed U.S. Environmental Protection Agency (EPA) the first real statutory power at the Federal level to control air pollution.

As mentioned in Chapter 2, the EPA has established primary and secondary standards for each of five industrial air pollutants: particulates, sulfur dioxide, nitrogen oxides, hydrocarbons, and carbon monoxide. According to the Clean Air Act, primary levels —those necessary to protect human health—were to be achieved everywhere in the U.S. by May 31, 1975. Secondary standards—to protect vegetation and "public welfare"—were to be achieved within a "reasonable time." The Clean Air Act further authorized the EPA to promulgate emission limitation standards for extremely toxic air pollutants; however, only three such limits—for asbestos, beryllium and mercury—have been issued to date.

As the first step toward enforcing the primary and secondary standards, the EPA and the states divided the nation into approximately 250 air quality control regions (AQCR's). These regions consist of contiguous geographical areas which experience similar and shared pollution problems. The states then prepared "implementation plans," which they were supposed to submit by January, 1972, outlining the enforcement steps they would take to bring air pollution levels in each of the AQCR's into compliance with the standards. There have been numerous disputes and even court battles over state plan provisions, but by 1974 most sections of most plans had received EPA approval, and, after some delay, are now being put into effect.

A thorough discussion of the Clean Air Act Amendments, state implementation plans, and air pollution standards appears in *A Citizen's Guide to Clean Air*, published by the Conservation Foundation (Washington, D.C., free).

State Air Agency Emission Inventories

The Clean Air Act contained requirements as to what implementation plans must include. One of the most important of these requirements, to an industrial pollution researcher, is contained in Section 110. This part of the Act orders the states to provide for installation of monitoring equipment at all industrial pollution sources, for periodic reporting of emissions to the state pollution control agency, and for disclosure of these data to the public "at reasonable times." All state air pollution control agencies (see list) have instituted emission inventory programs to comply with Section 110. As a result, industrial emissions figures for the criteria air pollutants, and in some cases for toxic pollutants as well, should now be available.

"Point" sources of air pollution: industrial smokestacks in Birmingham, Alabama.

STATE AIR POLLUTION CONTROL AGENCIES

ALABAMA
State of Alabama Department of
  Public Health
State Office Building
Montgomery, Alabama 36104

ALASKA
State of Alaska
Department of Environmental
  Conservation
Pouch O
Juneau, Alaska 99801

ARIZONA
Division of Air Pollution Control
4019 N. 33rd Avenue
Phoenix, Arizona 85017

ARKANSAS
Arkansas Department of Pollution
  Control and Ecology
8001 National Drive
Little Rock, Arkansas 72209

CALIFORNIA
Air Resources Board
1025 P Street
Sacramento, California 95814

COLORADO
Air Pollution Control Division
Colorado Department of Health
4210 East 11th Avenue
Denver, Colorado 80220

CONNECTICUT
Air Compliance Section
Department of Environmental
  Protection
165 Capitol Avenue
Hartford, Connecticut 06115

DELAWARE
Delaware Department of Natural
  Resources and Environmental
  Control
Division of Environmental Control
Tatnall Building, Capitol Complex
Dover, Delaware 19901

DISTRICT OF COLUMBIA
District of Columbia Department of
  Environmental Services
Bureau of Air and Water Pollution
  Control
25 K Street, N.E.
Washington, D.C. 20002

FLORIDA
Department of Pollution Control
Tallahassee Bank Building, Suite 300
315 South Calhoun Street
Tallahassee, Florida 32301

GEORGIA
Environmental Protection Division
Department of Natural Resources
47 Trinity Avenue, S.W.
Atlanta, Georgia 30334

HAWAII
Air Sanitation Branch
Division of Environmental Health
1250 Punchbowl Street
Honolulu, Hawaii 96813

IDAHO
Department of Environmental
  Protection and Health
Air Pollution Control Section
Statehouse
Boise, Idaho 83707

ILLINOIS
Environmental Protection Agency
2200 Churchill Road
Springfield, Illinois 62706

INDIANA
Indiana State Board of Health
1330 West Michigan Street
Indianapolis, Indiana 46206

IOWA
Environmental Engineering Service
Iowa State Department of Health
Lucas State Office Building
Des Moines, Iowa 50319

KANSAS
Kansas State Department of Health
535 Kansas Avenue
Topeka, Kansas 66603

KENTUCKY
Kentucky Air Pollution Control
   Commission
275 East Main Street
Frankfort, Kentucky 40601

LOUISIANA
Air Control Section
Bureau of Environmental Health
Louisiana State Department of Health
P.O. Box 60630
New Orleans, Louisiana 70160

MAINE
Department of Environmental
   Protection
Bureau of Air Pollution Control
State House
Augusta, Maine 04330

MARYLAND
Bureau of Air Quality Control
Maryland State Department of Health
   and Mental Hygiene
610 North Howard Street
Baltimore, Maryland 21201

MASSACHUSETTS
Bureau of Air Quality Control
Division of Environmental Health
Department of Public Health
600 Washington Street
Boston, Massachusetts 02111

MICHIGAN
Division of Air Pollution Control
Michigan Department of Public Health
3500 North Logan Street
Lansing, Michigan 48914

MINNESOTA
Division of Air Quality
Minnesota Pollution Control Agency
717 Delaware Street, S.E.
Minneapolis, Minnesota 55440

MISSISSIPPI
Mississippi Air and Water Pollution
   Control Commission
Robert E. Lee Building
Jackson, Mississippi 39205

MISSOURI
Missouri Air Conservation Commission
P.O. Box 1062
117 Commerce Avenue
Jefferson City, Mo. 65101

MONTANA
Montana State Department of Health
   and Environmental Sciences
Cogswell Building
Helena, Montana 59601

NEBRASKA
Division of Air Pollution Control
State Department of Environmental
   Control
P.O. Box 94653
State House
Lincoln, Nebraska 68509

NEVADA
Bureau of Environmental Health
201 South Fall Street
Carson City, Nevada 89701

NEW HAMPSHIRE
New Hampshire Air Pollution
   Control Agency
61 South Spring Street
Concord, New Hampshire 03301

NEW JERSEY
New Jersey State Bureau of Air
   Pollution Control
Division of Environmental Quality
Department of Environmental
   Protection
P.O. Box 1390
Trenton, New Jersey 08625

NEW MEXICO
Environmental Improvement Agency
PERA Building
College & West Manhattan
Santa Fe, New Mexico 87501

NEW YORK
New York State Department of
   Environmental Conservation
50 Wolf Road
Albany, New York 12201

NORTH CAROLINA
Department of Natural and Economic
   Resources
Office of Water and Air Resources
P.O. Box 27687
Raleigh, North Carolina 27611

NORTH DAKOTA
North Dakota State Department of
   Health
State Capitol
Bismarck, North Dakota 58501

OHIO
Air Pollution Unit
Ohio Department of Health
450 East Town Street
Columbus, Ohio 43216

OKLAHOMA
Air Pollution Control Division..
Environmental Health Services
Oklahoma State Department of Health
3400 North Eastern Avenue
Oklahoma City, Okla. 73105

OREGON
Department of Environmental Quality
1234 S.W. Morrison Street
Portland, Oregon 97205

PENNSYLVANIA
Bureau of Air Quality and Noise
  Control
Department of Environmental
  Resources
Commonwealth of Pennsylvania
P.O. Box 2351
Harrisburg, Pennsylvania 17105

RHODE ISLAND
Rhode Island Division of Air Pollution
  Control
204 Health Building
Davis Street
Providence, Rhode Island 02908

SOUTH CAROLINA
South Carolina Pollution Control
  Authority
Division of Air Pollution Control
1321 Lady Street
P.O. Box 11628
Columbia, S. C. 29211

SOUTH DAKOTA
South Dakota State Department of
  Health
Division of Sanitary Engineering and
  Environmental Protection
Air Quality Control Program
Office Building #2
Pierre, South Dakota 57501

TENNESSEE
Division of Air Pollution Control
Tennessee Department of Public
  Health
C2-212 Cordell Hull Building
Nashville, Tennessee 37219

TEXAS
Air Pollution Control Services
Texas State Department of Health
1100 West 49th Street
Austin, Texas 78756

UTAH
Utah State Division of Health
44 Medical Drive
Salt Lake City, Utah 84113

VERMONT
Agency of Environmental Conservation
Air Pollution Control
P.O. Box 489
Montpelier, Vermont 05602

VIRGINIA
State Air Pollution Control Board
Room 1106
Ninth Street Office Building
Richmond, Virginia 23219

WASHINGTON
Washington State Department of
  Ecology
P.O. Box 829
Olympia, Washington 98504

WEST VIRGINIA
West Virginia Air Pollution Control
  Commission
1558 Washington Street East
Charleston, West Virginia 25311

WISCONSIN
Wisconsin Department of Natural
  Resources
Bureau of Air Pollution and Solid
  Waste Disposal
Box 450
Madison, Wisconsin 53701

WYOMING
Air Quality Section
Wyoming Division of Health and
  Medical Services
State Office Building
Cheyenne, Wyoming 82001

You should be able to obtain the inventory data for the factory you are investigating by sending a letter of request to the director of the agency in the state where the factory is located. The letter should ask for the following specific facts concerning each criteria pollutant:

(1) Total factory emissions for each of the past three years.

(2) A breakdown of the emissions by source within the factory for each of the past three years.

(3) Factory emissions allowable under law, by source.

(4) Total present emissions from all sources within the city, county or air basin where the factory is located.

To expedite your request, it is good practice to enclose a chart on which data can be easily inserted (see sample chart).

If the response you receive is incomplete, or if, as is sometimes the case, the state air agency has a policy of not answering mail requests for emissions statistics, you will probably have to visit the agency's office in person to obtain the data. If you come to an interview with an agency official prepared with your chart listing the factory's major production facilities and the statistics you are seeking, the official will know exactly what you need.

Two further obstacles may be encountered in securing data. First, despite the Clean Air Act mandate, emission data may not have been collected. As of March, 1974, for example, New Jersey still had not gathered data on hydrocarbon or carbon monoxide emissions from any of the five huge petroleum refineries within the state. Should this problem arise in the case of the factory you are analyzing, the only real recourse is to pressure the state pollution control agency and the EPA to enforce the law. You may be able to get a rough estimate of emissions from a co-operative agency official prior to completion of precise emission monitoring.

The other potential problem is that the state agency may be reluctant to release emission data. If the official you initially contact gives a negative response, try contacting another official at a later date, as he or she may be more sympathetic. In the face of continued difficulty, you should make clear that Section 110 of

70

Sample air data chart.

EMISSIONS INVENTORY
(lbs/hour)

COMPANY:
FACILITY:
DATE:

Pollutant	Source (name of production process)	Allowable Emissions	1974 Emissions	1973 Emissions	1972 Emissions	Total......... (city, county air, basin) emissions
Sulfur Dioxide						
Particulates						
Nitrogen Oxides						
Carbon Monoxide						
Hydrocarbons						

the Clean Air Act requires public disclosure of emissions data. If all else fails, legal proceedings can be instituted against the state agency. However, before taking such a step, you should consult the U.S. EPA, which may also be able to provide emissions data.

Federal National Air Data Bank

In 1971, the EPA began setting up a National Air Data Bank in Durham, North Carolina, to catalogue all U.S. sources of air pollution. After three years, this program is well along the road to being operational. As of July, 1974, emission statistics for the criteria pollutants had been collected for 85,000 "point" emission sources, *i.e.*, smokestacks, at 30,000 factories. Eventually information will be on file for all point sources discharging more than 100 tons of a criteria pollutant a year. The statistics, which come from emission inventories prepared by state air agencies or companies themselves, are updated every six months. Emissions from "mobile sources" of pollution, primarily automobiles, in 3,100 counties throughout the country, are also currently included in this data bank. Sometime in 1975, the EPA will expand the National Air Data Bank to include figures on toxic pollutant release.

Particular factory statistics can be requested from the Data Processing Center of the Monitoring and Data Analysis Division in Durham, N. C., or from a regional EPA office (see list, p. 83), which is sent computer print-outs every six months listing point source emissions within the region. The Durham office also publishes "National Emission Reports," which indicate total emissions of various air pollutants in each Air Quality Control Region.

Fugitive Emission Data Sources

Some sizable industrial air pollution emissions are not reflected in emission inventories, either because they are intermittent or because they are too difficult to measure. These include "fugitive dust" and gases, which can emanate from windows, or seep from cracks in furnace or oven doors, and the intermittent emissions which can occur when raw materials are added to processes.

At a U.S. Steel mill in Pittsburgh, for example, air pollution control officials found that the emissions from the smokestack atop an electric furnace met with county regulations, but that smoke exceeding legal limits was bursting through the windows and roof of the furnace shop itself. The company was eventually required to install $5 million worth of exhaust fans and other equipment to suck the polluted air into a filter system and send it out the stack.

Other emissions, such as dust from raw material stockpiles (which varies with wind conditions) are unquantified, simply because they are unquantifiable. Rough approximations of such emissions can sometimes be made with the help of an engineer. Again, an interview at a state or local air agency is the best source of information.

"Fugitive" or intermittent air emissions usually affect the environment at the immediate plant site more than they do the region's general air quality. They tend to hover at or near the ground in and outside the plant, until they settle out or decompose. Therefore, the Department of Labor and labor unions, the two groups most concerned with the quality of the workplace environment, tend to be concerned about and keep track of such problems. Some labor union locals have been collecting information about air pollution levels near work sites for years, trying to determine where better ventilation should be installed, whether workers should wear respirators, and where workers should not be allowed to work at all. The Occupational Safety and Health Administration (OSHA), set up within the Labor Department in 1970 to enforce workplace health and safety standards, has monitored air quality at thousands of industrial operations.

Both unions and the OSHA may have data on the factory you are analyzing. OSHA factory inspection reports are available at the agency's Area Offices, located in each state (see list). However, advance notice of your desire to view the reports is required, as agency officials are required to screen out any information on the company considered proprietary before making them available to the public. Some states have health and safety agencies or local boards of health which might be able to supply additional figures or viewpoints.

The company operating the factory may also have surveyed and measured workplace air pollution and recorded rates at which "fugitive" emissions escape and diffuse into the external environ-

Fugitive emissions from coal piles near Charleston, West Virginia. The coal is burned at an electric power plant.

OCCUPATIONAL SAFETY AND HEALTH ADMINISTRATION
AREA OFFICES

ALABAMA

Harold J. Montegue, Area Director
Commerce Building—Room 600
118 North Royal Street
Mobile, Alabama 36602

G. Larry Wyatt, Area Director
Todd Mall
2047 Canyon Road
Birmingham, Alabama 35216

ALASKA

Darrell Miller, Area Director
Federal Building—Room 227
605 West Fourth Avenue
Anchorage, Alaska 99501

ARIZONA

Lawrence E. Gromachey, Area Director
Amerco Towers—Suite 910
2721 North Central Avenue
Phoenix, Arizona 85004

ARKANSAS

Robert A. Griffin, Area Director
Donaghey Building—Room 303
103 East 7th Street
Little Rock, Arkansas 72201

CALIFORNIA

Edward Jones, Acting Area Director
100 McAllister Street
Room 1706
San Francisco, California 94102

Bernard L. Tibbetts, Area Director
Hartwell Building—Room 401
19 Pine Avenue
Long Beach, California 90802

COLORADO

Jerome J. Williams, Area Director
Squire Plaza Building
8527 W. Colfax Avenue
Lakewood, Colorado 80215

CONNECTICUT

Harold R. Smith, Area Director
Federal Building—Room 617 B
450 Main Street
Hartford, Connecticut 06103

DELAWARE

DISTRICT OF COLUMBIA

FLORIDA

George D. Barlow, Area Director
Bridge Building—Room 204
3200 East Oakland Park Blvd.
Fort Lauderdale, Florida 33308

William W. Gordon, Area Director
2809 Art Museum Drive
Suite 4
Jacksonville, Florida 32207

Area Director:
Suite 200
1802 North Trask Street
Tampa, Florida 33609

GEORGIA

A. de Jean King, Area Director
Riverside Plaza Shopping Center
2720 Riverside Drive
Macon, Georgia

Joseph L. Camp, Area Director
Building 10—Suite 33
LaVista Perimeter Park
Tucker, Georgia 30384

Bernard E. Addy
Enterprise Building
Suite 204
6605 Abercorn St.
Savannah, Ga. 31405

HAWAII

Paul F. Haygood, Area Director
333 Queen Street—Suite 505
Honolulu, Hawaii 96813

IDAHO

Richard Jackson, Area Director
228 Idaho Building
216 North 8th Street
Boise, Idaho 83702

ILLINOIS

William Funcheon, Area Director
300 South Wacker Drive
Room 1200
Chicago, Illinois 60606

INDIANA

J. Fred Keppler, Area Director
U.S. Post Office and Courthouse
Room 423
46 East Ohio Street
Indianapolis, Indiana 46204

IOWA

Area Director:
210 Walnut Street—Room 643
Des Moines, Iowa 50309

KANSAS

Roger A. Clark, Area Director
Petroleum Building—Suite 312
221 South Broadway
Wichita, Kansas 67202

KENTUCKY

Frank P. Flanagan, Area Director
Suite 554-E
600 Federal Plaza
Louisville, Kentucky 40202

LOUISIANA

J. E. Powell, Area Director
Fourth Floor
546 Carondelet Street
New Orleans, Louisiana 70130

MAINE

MARYLAND

Maurice R. Daly, Area Director
Federal Building—Room 1110 A
31 Hopkins Plaza
Charles Center
Baltimore, Maryland 21201

MASSACHUSETTS

John V. Fiatrone, Area Director
Room 703
Custom House Building
State Street
Boston, Massachusetts 02109
Rudolph Bayerle, Jr., Area Director
U.S. Post Office and Courthouse
436 Dwight Street
Room 501
Springfield, Mass. 01103

MICHIGAN

G. Cunningham, Acting Area Director
Michigan Theater Building
220 Bagley Avenue—Room 626
Detroit, Michigan 48226

MINNESOTA

Vernon Fern, Area Director
110 South Fourth Street
Room 437
Minneapolis, Minn. 55401

MISSISSIPPI

James E. Blount, Area Director
5670 I—55 North Frontage Rd. East
Jackson, Mississippi 39211

MISSOURI

Robert J. Borchardt, Area Director
1627 Main Street—Room 1100
Kansas City, Missouri 64108
Angelo F. Castranova, Area Director
210 North 12th Blvd.—Room 554
St. Louis, Missouri 63101

MONTANA

Area Director:
Petroleum Bldg.—Suite 525
2812 First Avenue North
Billings, Montana 59102

NEBRASKA

Warren P. Wright, Area Director
City National Bank Building
Room 803
Harney and 16th Streets
Omaha, Nebraska 68102
Oscar F. DiSilvestro, Area Director
113 West 6th Street
North Platte, Nebraska 69101

NEVADA

Ivan Schulenburg, Area Director
1100 E. William Street
Carson City, Nevada 89701

NEW HAMPSHIRE

Francis R. Amirault
Federal Building—Room 425
55 Pleasant Street
Concord, New Hampshire 03301

NEW JERSEY

William J. Dreeland, Area Director
Federal Office Building
970 Broad Street
Room 1435 C
Newark, New Jersey 07012

NEW MEXICO

Area Director
Room 302
Federal Building
421 Gold Avenue, S.W.
P.O. Box 1428
Albuquerque, N.M. 87103

NEW YORK

Nicholas A. DiArchangel
Area Director
90 Church Street
Room 1405
New York, N. Y. 10007

Chester Whiteside, Area Director
Room 203
Midtown Plaza
700 East Water Street
Syracuse, N. Y. 13210

James H. Epps, Area Director
370 Old Country Road
Garden City, N. Y. 11530

NORTH CAROLINA

Quinton F. Haskins, Area Director
Federal Office Building—Room 406
310 New Bern Avenue
Raleigh, North Carolina 27601

NORTH DAKOTA

OHIO

Peter Schmitt, Area Director
360 S. Third Street—Room 109
Columbus, Ohio 43215

Kenneth Bowman, Area Director
847 Federal Office Building
1240 East Ninth Street
Cleveland, Ohio 44199

Ronald McCann, Area Director
Federal Office Building—Room 5522
550 Main Street
Cincinnati, Ohio 45202

Glenn Butler, Area Director
Federal Office Building—Room 734
234 N. Summitt Street
Toledo, Ohio 43604

OKLAHOMA

James T. Knorpp, Area Director
Petroleum Building—Room 512
420 South Boulder
Tulsa, Oklahoma 74103

OREGON

Eugene Harrower, Area Director
Pittock Block—Room 526
921 S.W. Washington Street
Portland, Oregon 97205

PENNSYLVANIA

Harry Sachkar, Area Director
Federal Building—Room 4456
600 Arch Street
Philadelphia, Pennsylvania 19106

Harry G. Lacey, Area Director
Jonnet Building—Room 802
4099 William Penn Highway
Monroeville, Pennsylvania 15146

RHODE ISLAND

SOUTH CAROLINA

William N. Duncan, Area Director
1710 Gervais Street—Room 205
Columbia, South Carolina 29201

TENNESSEE

Eugene E. Light, Area Director
1600 Hayes Street—Room 302
Nashville, Tennessee 73203

TEXAS

Charles J. Adams, Area Director
Adolphus Tower—Suite 1820
1412 Main Street
Dallas, Texas 75202

Robert B. Simmons, Area Director
Room 421, Federal Building
1205 Texas Avenue
Lubbock, Texas 79401

Thomas T. Curry, Area Director
2320 LaBranch—Room 2118
Houston, Texas 77004

Herbert M. Kurtz, Area Director
Room 215, Jackson Keller Road
San Antonio, Texas 78213

UTAH

Charles F. Hines, Area Director
Executive Building—Suite 309
455 East 4th Street
Salt Lake City, Utah 84111

VERMONT

VIRGINIA

Lapsey C. Ewing, Jr., Area Director
Federal Building—Room 8081
P.O. Box 10186
400 North 8th Street
Richmond, Virginia 23240

WASHINGTON

Richard L. Beeston, Area Director
121-107th Street, N.E.
Bellevue, Washington 98004

WEST VIRGINIA

Byron R. Chadwick, Area Director
Charleston National Plaza—Room 1726
Charleston, West Virginia 25301

WISCONSIN

Robert B. Hanna, Area Director
Clark Building—Room 400
633 W. Wisconsin Avenue
Milwaukee, Wisconsin 53203

WYOMING

ment. It may be willing to review its findings with you. Some findings may also have been entered as evidence in hearings or court cases. Kennecott Copper, for example, provided data on a copper smelter's fugitive sulfur dioxide at a hearing on the tightening of air standards in New Mexico. The company was maintaining that on the basis of its fugitive emissions alone, regardless of how good its stack gas controls were, it would be incapable of meeting the proposed limit. (The standard was not strengthened.)

WATER POLLUTION DATA

The Refuse Act of 1899 and the Water Pollution Control Act Amendments of 1972, which together form the backbone of the Federal water pollution abatement program, have caused voluminous data about industrial water pollution discharges to be gathered and made publicly available.

The Law: The Refuse Act

The 1899 Rivers and Harbors Act, also known as the Refuse Act, which lay dormant for 71 years until it was reactivated by the House Subcommittee on Conservation and Natural Resources in 1970, said simply that, "it shall not be lawful to throw . . . any refuse matter of any kind . . . into any navigable water of the United States," without a permit to do so from the U.S. Army Corps of Engineers. Violators of the Act were subject to a $25,000-a-day fine, half of which was offered as a bounty to anyone who provided the Department of Justice with evidence of a violation. As of 1970, virtually every factory discharging waste water was in violation of this law as none had ever filed for a permit.

The revived Refuse Act was intended to constitute a new tool for enforcing water pollution standards, which had been languishing because of weaknesses in the much-amended Water Pollution Control Act of 1956. On December 23, 1970, the Corps was directed by Executive Order to initiate a permit program. It required industrial polluters to submit permit applications by

July 1, 1971, for each factory which discharged water into a navigable waterway. However, the Corps was not to issue any permits authorizing discharges of any pollutant in excess of existing Federal or state regulations. A polluter wishing to avoid prosecution for not having a permit would thus have to reduce its pollution loads to acceptable levels.

The Refuse Act permit program remained in effect for nearly two years. By October 12, 1972, companies had filed applications for 28,000 permits with the Corps. Only about 20 permits were ever granted, and many applications were never properly completed. However, the permit program was subsequently incorporated, more or less in total, into the new National Pollution Discharge Elimination System (NPDES) permit program set up under the 1972 Water Pollution Control Act. Any company which had filed a permit application under the old program was not required to refile a new one. As a result, Refuse Act Permit Program (RAPP) applications remain one of the most important sources of information about industrial water pollution control. Both RAPP and NPDES applications from factories in each EPA region are available for public examination at the EPA regional office, as described below.

RAPP Applications

A Refuse Act Permit application, properly filled in, can tell you a great deal about a factory's water pollution. However, reading the form can be a complicated task. A company instruction manual, prepared in 1971 by the U.S. Army Corps of Engineers to help applicants complete the forms correctly, can be a great help in understanding a RAPP application. This booklet, entitled *Permits for Work and Structures in, and for Discharges into Navigable Waters* is available free from the Corps office in Washington, D.C., or from any EPA regional office.

Section I of a RAPP application, three pages long, gives background information about the plant, including its location, number of employees, and the name of the water bodies into which it discharges effluent. It also lists the amount of water taken into the facility each day, indicates how it is used (*i.e.*, in cooling or in industrial processes), and what happens to the water not discharged back into the waterway (*i.e.*, evaporative losses, pumping, of contaminated waste water into underground deep wells, etc.).

Although the RAPP form does not require information on the total flow or size of the water body from which the factory draws, this data can be obtained from the state office (usually located in the state capital) of the U.S. Department of the Interior, U.S. Geological Survey. An annual report entitled *Water Resources Data for [name of state], Part I: Surface Water Records* (free, on request to the District Chief, Water Resources Division, U.S. Geological Survey in the state) notes the daily flows at various points of all major waterways in the state.

Section II of a RAPP application contains data on the amount of pollution the plant emits. A complete 12-page Section II must be filled out for each "outfall," the term commonly used to describe any pipe discharging waste water. Some plants have only one "outfall," and thus must file just one "Section II"; others can have dozens.

In a Section II the applicant is required to report the quantity of water discharged, and the temperature and pH of both the intake water and discharge water at that outfall. The section also requires data on discharges of numerous specific polluting substances, in terms of the average concentration of the pollutant in the intake, the average and maximum concentrations in the discharge water, and the average and maximum pounds of the pollutant discharged each day from the outfall.

Although space is provided for information on about 50 specific pollutants, companies need not report emissions for all of them. The Corps established lists of what it considered the principal pollutants from each major industry, and applicants need to describe only those discharges. Companies in the sugar refining industry, for example, must report on emissions of only three pollutants, while applicants in the organic chemical industry must report on emissions of 26. The lists, indexed by SIC code numbers, appear in the Corps instruction manual mentioned above.

A potentially confusing aspect of a RAPP application is that it sometimes fails to indicate the units in which discharges are reported. The Corps instruction manual, however, will tell you what units are used. In general, intake and discharge concentrations are reported in parts per million (ppm), except for metals, which are in parts per billion (ppb).* Average daily discharges are reported in pounds per day; water flows appear either in gallons per minute (GPM) or million gallons per day (MGD).

A Section II will also include data on sources of waste water, and on treatment systems. Item 11 requires a list of the production operations discharging water to the outfall pipe. Item 21 requires a description of the pollution abatement equipment through which the effluent passes before being discharged. The Corps has developed a code word for each of 73 standard waste water treatment practices. For example, PSEDIM stands for a sedimentation tank and BAERAT for an aerated lagoon. RAPP applicants may include full descriptions of factory treatment systems or just list the code words characterizing operating pollution controls (see chart for key). The code is fully explained in the Corps instruction manual.

Unfortunately, some RAPP applications may be incomplete, or contain outdated data, inconsistencies or even errors, as the EPA and the Corps have sometimes been less than diligent in requiring full compliance with application instructions. If missing data cannot be calculated from information supplied elsewhere in the form, and does not appear in correspondence appended to the application, it may be necessary to prod the EPA to obtain the data from the applicant. A Council on Economic Priorities researcher investigating the steel industry, for example, once read an application which omitted all discharge data for five of 21 outfall pipes known to exist at a large mill. When, after repeated requests, the EPA finally obtained the missing information, the five outfalls turned out to be discharging a daily total of 22 million gallons of water, laden with over 3,000 pounds of suspended solids and nearly 100 pounds of oil.

---

* One part per million (ppm) equals one milligram per liter (mg/l) or .0000083 pounds per gallon of water.

COMMONLY USED CODE WORDS FOR WASTE WATER TREATMENT SYSTEMS APPEARING IN RAPP AND NPDES APPLICATION FORMS

ESEPAR	separate drainage systems
EMERGE	emergency storage facilities
EPUMPS	use of pumps and valves with special seals
DCHEMI	chemical regeneration
RUSEOR	use or sale of wastes
RHEATR	heat recovery
LEVAPO	evaporation or incineration of wastes
PEQUAL	equalization
PSCREE	screening
PAERAT	pre-aeration
PSEDIM	sedimentation
PFLOAT	flotation
CNEUTR	neutralization
COOAGU	chemical coagulation
BACTIV	activated sludge
BTRICK	trickling filter
BAERAT	aerated lagoon
STHICK	thickening
SLAGOO	sludge lagoon
SLANDD	land disposal
WDEEPW	deep well disposal
WDISCH	surface discharge

As noted above, a factory's RAPP form should be on file at the EPA regional office for the region in which the factory is located (see list). The application should be available for inspection in person at no charge, or obtainable by mail at a copying cost of $.20 a page. Before ordering a copy, however, you may want to check the form's length over the phone. If the factory you are analyzing has dozens of outfall pipes, as some do, it can have an application of up to a thousand pages, for which the xeroxing bill would be hundreds of dollars. A trip to the regional office may prove more economical.

The Law: The Water Pollution Control Act

On October 12, 1972, the Refuse Act permit program was pre-empted by passage of the Federal Water Pollution Control Act Amendments. This law did to the Federal water pollution control program what the Clean Air Act did to national air pollution control: it gave a centralized Federal agency power to set and enforce national water pollution standards. The EPA was given the responsibility of preparing a set of discharge standards, called "effluent guidelines," for every major U.S. industry. The states were given the job, subject to EPA review, of developing pollution concentration limits, called "water quality standards," for each of their navigable waterways.

EPA's effluent guidelines, discussed in their "proposed" form in the *Development Documents* noted in Chapter 2 and in their "final" form in the *Federal Register* of the date on which they were promulgated, precisely define the degree of pollution control which various industries must achieve within two different time periods.

The first standard represents the level of pollution control obtainable by the "best practicable" treatment systems on the market, a term which is defined as including cost-benefit considerations. These standards are commonly called "BPT" (best practicable technology) or Level I guidelines. The EPA and the state regulatory agencies were directed by the 1972 Act to develop and enforce water pollution abatement strategies designed to guarantee that industrial discharges will comply with BPT standards by July 1, 1977.

ENVIRONMENTAL PROTECTION AGENCY REGIONAL OFFICES

Region	States Included	Address
I	Connecticut, Maine, Massachusetts, New Hampshire, Rhode Island, Vermont	John F. Kennedy Federal Building Boston, Mass. 02203
II	New Jersey, New York, Puerto Rico, Virgin Islands	26 Federal Plaza New York, New York 10007
III	Delaware, District of Columbia, Maryland, Pennsylvania, Virginia, West Virginia	Curtis Building 6th and Walnut Streets Philadelphia, Pennsylvania 19106
IV	Alabama, Florida, Georgia, Mississippi, Kentucky, North Carolina, South Carolina, Tennessee	1421 Peachtree Street, N.E. Atlanta, Georgia 30309
V	Illinois, Indiana, Minnesota, Michigan, Ohio, Wisconsin	Post Office Building 433 W. Van Buren Street Chicago, Illinois 60607
VI	Arkansas, Louisiana, New Mexico, Oklahoma, Texas	1600 Patterson Suite 1100 Dallas, Texas 75201
VII	Iowa, Kansas, Missouri, Nebraska	1735 Baltimore Avenue Kansas City, Missouri 64108
VIII	Colorado, Montana, North Dakota, South Dakota, Utah, Wyoming	1860 Lincoln Street Denver, Colorado 80203
IX	Arizona, California, Hawaii, Nevada, Guam, American Samoa	100 California Street San Francisco, California 94111
X	Washington, Oregon, Idaho, Alaska	1200 Sixth Avenue Seattle, Washington 98101

"Outfall" pipes at Union Carbide plant on the Kanawha River, near South Charleston, West Virginia.

ond standard for each industry is stricter, reflecting the
of pollution control obtainable by use of the "best avail-
ollution control systems. Companies are required to meet
iidelines, called "BAT" (best available technology) or
standards, by July 1, 1983.

was also ordered to set "NSPS" or New Source Per-
Standard guidelines, regulating the amount of pollution
from new factories. NSPS or Level III guidelines take
nt processes as well as treatment equipment which can
l in factory design to help reduce or prevent pollution.
lelines, fully discussed in Chapter 6, are at least as
AT" Level II limits.

State water quality standards specifically define the levels of
purity which each waterway in the state must achieve within
similar time periods. All navigable waterways are to be upgraded
to a level which will permit propagation of fish and wildlife and
allow recreation *on* the water, *i.e.* boating, by 1977, and to a level
which will safely permit recreation *in* the water, *i.e.* swimming,
by 1983.

The Water Pollution Control Act set as a final goal complete
elimination of all industrial water pollutant discharges by July
1, 1985. In practical terms, this means total reliance on waste
water recycling, known as "closed loop" systems.

All of the Act's provisions are thoroughly discussed in a very
readable, informative booklet published by the Izaak Walton
League entitled *A Citizen's Guide to Clean Water* (Arlington,
Va., free).

NPDES Applications

To realize its goals, the Water Pollution Control Act replaced the
Refuse Act permit program with the National Pollution Discharge
Elimination System (NPDES). Late in 1972 all RAPP applications were transferred from the Corps of Engineers to the EPA
and to state pollution control agencies, and became NPDES
applications. Under NPDES, not only industrial, but municipal

and agricultural dischargers as well (RAPPs applied only to the former), must file for a water discharge permit similar to a Refuse Act permit. The deadline for filing was April 16, 1973; however, companies which had already filed for a Refuse Act permit were allowed simply to resubmit that application.

The inclusion of municipal sewage treatment plants means, among other things, that companies which divert their waste water to these treatment facilities rather than discharge it directly into a navigable waterway no longer can avoid having their discharges reported to the government, and regulated. A municipal sewage treatment plant handling industrial wastes must indicate in its permit application the level of pollution in *incoming* water. If the industrial effluent is polluted with chemicals that the treatment plant cannot remove, the industry is required to pretreat the water before piping it on.

The EPA and the state pollution control agency use the statistics on a permit application to assess whether a factory's performance both meets EPA effluent guidelines and allows state water quality standards to be achieved. If its performance proves adequate, a permit of up to five years is granted. If its discharges exceed these levels, the EPA, the state, and the company develop a "compliance schedule" which, if properly followed by the company, will bring the factory's discharges into compliance with both industry guidelines and stream quality standards by July, 1977. Target dates for completion of various steps in the compliance schedule may be no more than nine months apart.

The company then receives a permit with its compliance schedule attached. The permit is valid only so long as the company meets the stipulated pollution reduction schedule.

Between 1977 and 1983, whenever their permits expire, companies will have to reapply. The state and the EPA will then scrutinize their factories' discharges anew against the stricter 1983 standards. If necessary, new compliance schedules will be developed to achieve further pollution reductions.

The NPDES permit program itself is described in an EPA publication called *Toward Cleaner Water* (GPO #546-312/140, $1.50), as well as in The Conservation Foundation's *Water Quality Training Institute Kit,* mentioned in Chapter 2.

At present, many factories have Refuse Act permit applications on record. If permits have been granted, they include, where necessary, compliance schedules developed under the new program. In all such cases, data about water pollution and pollution control can be obtained from the RAPP application as described above.

In the future, however, companies will file for discharge permits using the newly issued NPDES forms. These differ slightly from the RAPP forms. Section II of an NPDES requires disclosure of the maximum quantities of raw materials consumed in production processes and the precise maximum quantities of products produced. More data about the frequency of discharge and thermal pollution is also required.

However, Section II no longer requires the applicant to report the average or maximum *pounds* of pollutants discharged daily, only the average *concentration* of each pollutant in both the intake and discharge water.

There is also a new "Section III" under the NPDES program entitled "Waste Abatement Requirements and Implementation Schedule." Here the EPA describes the compliance schedule which the applicant must follow in order to keep a permit. Section III of the new form is used even where the first two sections consist of forms completed under the old Refuse Act program.

NPDES permits and applications are available for public inspection at regional EPA offices and at state pollution control agencies. Again, copies will be sent on request. However, while the standard NPDES application is shorter than the RAPP form, duplicating costs can still be very high.

Prior to ordering a full NPDES application, you can often obtain a fact sheet summarizing the contents of the application, or the permit, free of charge. If the permit application is for a discharge of more than 500,000 gallons of water a day, or one containing toxic substances, the EPA or the state agency must prepare such a fact sheet before issuing a permit. These sheets do not include production data, however, limiting their usefulness.

BACKGROUND DATA

While emission inventories and RAPP and NPDES applications contain enormously valuable statistics, they do not include all the information needed to evaluate a factory's pollution control performance and how it can be improved. Besides tabulating the exact amount of pollution emerging from a factory's smokestacks and discharge pipes, you also need to gather additional details about the factory's control systems, solid waste disposal practices, raw materials and energy use, and financial position.

You may already have acquired some of the additional necessary data while doing the research outlined in Chapter 2. You should be able to check its accuracy, and obtain the rest, by contacting the company operating the factory.

Preparing a Questionnaire

The process of assembling data on the variety of factors affecting present and future pollution control performance of the factory you are investigating, can be greatly facilitated by using a questionnaire. This can function as a guide and an organizing tool, and serve as a general reminder of the information you seek. Referring to a questionnaire can save time and energy during interviews, and can keep you from repeating or forgetting to ask important questions.

You can use the sample company questionnaire shown opposite as a general blueprint for one of your own. It blocks out seven categories of background information that should be obtained. You will probably want to modify and adapt this format to include some new questions and to suit the type of facility you are analyzing.

The questionnaire begins by asking for basic data on factory production processes. Each major production unit should be listed in the left-hand column. Next to each should be spaces for noting the unit's age (year built), capacity, and production for the last year, or for any one "base" year. (It is important in your

SAMPLE COMPANY QUESTIONNAIRE

Production Processes

ase list the following for each production unit.

Unit (name)	Year Built	Capacity (ton/day)	197 Production
_____	_____	_____	_____
_____	_____	_____	_____
_____	_____	_____	_____

Air Pollution Controls

roduction Unit (name)	Type of Control	Design/ Tested Efficiency	Date Built	Capital Cost	Annual Operating Cost
_____	_____	_____ _____	_____	_____	_____
_____	_____	_____ _____	_____	_____	_____
_____	_____	_____ _____	_____	_____	_____

Water Pollution Controls

roduction Unit (name)	Type of Control	Design/ Tested Efficiency	Date Built	Capital Cost	Annual Operating Cost
_____	_____	_____ _____	_____	_____	_____
_____	_____	_____ _____	_____	_____	_____
_____	_____	_____ _____	_____	_____	_____

Solid Waste Management

Type	Quantity	Disposal or Recycling Method (please describe)
_____	_____	_____
_____	_____	_____
_____	_____	_____

5. Raw Material and Energy Consumption

List materials presently used for industrial processes.

Type	Amount (tons/day)
____	____
____	____
____	____

List fuel use.

Type	Amount (per day)	Sulfur Content (%)	Ash Content (%)
coal	_____ tons	_____	_____
oil	_____ bbl	_____	_____
gas	_____ scf	_____	_____

List electricity use in 197 . . .

Amount (kwh/year) _____

% generated _____

% purchased _____ from _____ (name utility)

6. Economic Factors

Value of the plant $_____

Recent plant investment—please list:

Year	Total Capital Expenditures— All Purposes ($)	Portion of Capital Expenditures— Pollution Control ($)
1974	_____	_____
1973	_____	_____
1972	_____	_____
1971	_____	_____
1970	_____	_____

Number of employees _____

Value of payroll (annual total) $_____

Taxes paid (annual total)   local $_____

state $_____

. Future Plans

Describe any pollution control plans projected for the next five years such as addition of pollution control equipment, fuel changes, or process modifications. Please indicate the timetable for implementing these plans, the projected costs, and the reductions of air and water emissions anticipated.

Describe any pollution control research and development projects currently under way.

research that you choose a single "base" year—usually the one most recently completed—for which to gather all data on pollution, production, fuel use, etc., so that you have comparable statistics.)

The next two sections of the questionnaire deal with air and water pollution control systems respectively. At most large factories, pollution control dates back to the earliest plant equipment. Hundreds of plants installed simple dust collectors and rudimentary settling basins as long ago as the beginning of the century, generally to prevent escape of valuable raw materials. Plants began using pollution control equipment on a large scale for the sake of protecting the environment in the 1950s and 1960s.

There is room in these two sections of the questionnaire to list each piece of factory production equipment in the left-hand column, and to fill in for each in spaces on the right the type of pollution control system operating, its design and tested efficiency, its date of installation, its cost at the time (known as "capital" cost), and its annual operating cost. In these sections, you can add questions about corporate air and water pollution control efforts which appear relevant but do not fall into the above categories. For example: "Does the company have an emergency lagoon system to entrap discharge water in the event of a treatment plant malfunction?"

The fourth section of the questionnaire contains questions about solid waste management. You need to know the type of by-products which the factory operation generates. For example, it may produce "tailings," the rock wastes left over after the desired material has been taken from an ore; or "sludge," the thick liquid wastes left over from water pollution control systems. You should also ask the quantity of each produced, and how or where the company gets rid of it. You may want to include further questions on the content of wastes and handling and disposal procedures if it appears that they may contain especially toxic substances.

Section five includes questions about plant raw material and energy consumption. There are spaces for listing the type and amount of raw materials used in making the plant's product, the amount and sulfur content of coal, oil or gas burned in its power boilers, and the plant's electricity consumption (the amount gen-crated at the plant and the amount bought from an electric utility). In the event that large amounts of electricity are purchased by the factory you are investigating, it may be relevant to look into the utility's environmental impact, since by buying power the factory may simply have "transferred" its pollution load to another site and company. (A procedure for making this kind of analysis is described in Chapter 6, under "Energy Use.")

The sixth section of the questionnaire focuses on economic data, requesting the total value of the plant as an asset, and a five-year record of capital expenditures for production and other equipment and for pollution controls. This information can suggest the existing corporate commitment to abating pollution. For example, certain manufacturing industries are presently investing about 10 percent of annual capital expenditures in pollution control. Any company greatly exceeding the industry percentage at a plant is probably making an exceptional current effort. (Companies that have spent very little in recent years may be found to be negligent. On the other hand, those spending little may have implemented such good and advanced control systems years ago that hardly any recent expenditures have been necessary; or they could be controlling pollution through process changes which are not tallied under capital expenditures for pollution control.)

The questionnaire also includes spaces for data on the number of employees at the plant, the size (in dollars) of its payroll, and the amount of state and local taxes it has paid in recent years. This information should provide some idea of the plant's contribution to the economy of the community.

The seventh and final section of the questionnaire asks about the company's future plans for the factory, since existing pollution control equipment should be continuously upgraded and, if the plant is emitting pollution in excess of the law, additional control steps must be taken. Is the company planning to install additional control equipment? Make process modifications? Change fuels? Is it conducting any research projects or experimental or pilot programs to overcome currently hard-to-solve pollution problems?

Once you have drafted up the specific list of questions to be included on your factory questionnaire, you may want to go over this list with several sympathetic scientists, agency officials, experts from conservation groups or other private organizations, or corporate people, whom you have encountered in the course of background research on the industry's pollution problems and control methods. Such individuals often can give useful criticisms, suggesting additional questions you should ask and pointing out questions which may be ambiguous, misleading, inappropriate or superfluous.

Equipped with a carefully drafted questionnaire, you can begin the task of filling in the blanks. Without doubt, the best source of background information on the factory you are interested in will be the company, which operates it. A company often makes much pollution-related information publicly available in printed materials. Corporate managers and technical personnel are also sometimes willing to discuss plant pollution control efforts with outside investigators, particularly if they see that the researchers have done extensive background work on potential pollution problems and are well-versed in the types of control equipment or procedures the company might employ.

General Sources

A good way to begin assembling data is by examining publicly available company literature, most of which you obtained in doing the background research suggested in Chapter 2, for relevant facts. You should review company press releases, public relations materials, annual reports, and 10K and 8K forms for the past several years for data on production, types and costs of recently installed pollution control equipment, and future plans. In-house magazines, which many large corporations distribute to employees and stockholders, and special publications on corporate environmental protection efforts, are often among the more useful corporate materials available. Jones and Laughlin Steel Company, for example, issues a statistic-laden annual progress report on pollution control (although its competitor, Armco Steel, publishes a more poetic but considerably less factual booklet entitled ". . . *and tis my belief that every flower loves the air it breathes.*")

Another very useful source of company-prepared data is prospectuses for special tax-exempt "pollution control revenue bonds" which Federal tax legislation, passed in 1968, permits corporations to issue to raise capital for pollution control purposes. These bonds are not issued by the corporation itself, but rather by an "industrial development authority" set up by a local government. The corporation receives the money and pays the interest to the bondholders through the authority. So far about $5 billion worth of pollution control revenue bonds have been sold.

Prospectuses for these bonds must describe completely the proposed project. They often give a history of pollution control at the factory as well. You may want to check *Moody's Bond Record*, available at most business libraries, to make sure that you have obtained prospectuses for all such projects at the factory you are analyzing.

As mentioned earlier in this chapter, companies must often file not only air and water emission data, but also data on raw material use, types of pollution control equipment operating, production, number of employees, and even their future plans (where compliance schedules have been worked out), with state air pollution agencies and the regional EPA office. The materials you have obtained from these agencies may thus include more of the facts needed to fill in additional blanks in your questionnaire.

As noted in Chapter 2, corporate pollution control efforts may also be chronicled in some detail in industry trade journals and in local newspapers. While this information comes to you "second-hand," and, unlike material filed with government agencies, is not legally required to be correct, these sources can nevertheless often supply useful data which you can verify with the company later.

Some trade journals have a year-end review issue which includes a summary of important changes in the industry and an index to the past year's articles. You may want to use this issue to locate articles with information on the company and factory you are interested in. You may also want to skim through all issues of one or two important industry journals for the past several years.

The local newspaper's "morgue," where back issues are filed, can also turn up factory background information. The completion of a pollution control facility is usually accompanied by a good deal of public relations fanfare, some of which is reported in the press. Newspapers also report pollution lawsuits and the proceedings of pollution variance boards and control agencies. They occasionally run feature articles on major regional environmental problems.

In checking back issues, note the names of reporters who wrote the most useful stories. Many newspapers now have one or more employees specializing in environmental reporting. In an interview, such a reporter may be able to offer information and insights on the factory's record which he or she did not include in news articles, and suggest other good information sources.

A Company Interview

After combing company brochures, prospectuses, annual reports, 10K forms, and agency filings, and reviewing materials in trade magazines and local newspapers, you should be ready to arrange a company interview. Corporate officials respond in a wide variety of ways to public inquiries, depending on their personal reactions and on company policies. The director of public relations of Dow Chemical Corporation—a company whose philosophy is essentially one of open discussion—has been known to send a well-prepared researcher right on to a personal interview with the Chairman of the Board, and then invite him to an all-day press briefing on environmental problems delivered by the Director of Pollution Control. On the other hand, U.S. Steel Corporation officials, outside of the public relations department, usually refuse to be interviewed by the public. A reporter for the *Wall Street Journal* once complained that he had tried unsuccessfully to arrange a meeting with U.S. Steel's new Director of Environmental Control, whose office was across the street from his, for seven months.

You should initially address your interview request to the factory manager; however, you will probably be referred to someone in the company's public affairs department. The official you see may begin by asking a ·seemingly endless number of questions about why you want information on company operations and what you intend to do with it. Officials are wary of releasing data that they feel might not be thoroughly understood or that might result in adverse publicity. Try to describe your interests, concerns and goals as straightforwardly as possible. If you can argue convincingly the responsibility of the company to disclose information on problems affecting the community, to explain its environmental protection efforts, and to conduct truly open dialogue and debate, the official may either refer you to someone else in the company who can answer technical questions, or hold a copy of your questionnaire until answers have been assembled.

If you do have an opportunity to interview one or more knowledgeable individuals in the company, such as the environmental controls director, you should come to the meeting with as much data in hand as possible. Corporate officials are sometimes more willing to verify data than to dig it out of their files and records themselves. Be sure to ask the company to check any figures you have obtained from governmental sources, including emission inventories and RAPP and NPDES applications. While this material is legally required to be correct, and is usually some of the best information available on factory emissions in terms of detail and accuracy, it can nevertheless be out of date or contain mistakes or oversights. Ask for a description of recent changes in factory pollution controls which may not be reflected in data for the previous year. Bring up any inconsistencies you have noticed in data collected from various sources.

If local company officials refuse to grant a personal interview, or decline to answer most of your questions, you might try sending a copy of the questionnaire with a cover letter discussing the purposes of your research, your desire to verify and update information you have already obtained, and the importance of corporate interaction with the community, to the Chairman of the Board, or President, of the company. That official may, in turn, advise personnel at the factory to be helpful, despite their initial reluctance.

A final important source of background data on the company's pollution control history and practices is the factory's labor union. Although the union probably maintains no records on the subject, employees who have worked at the plant for the last decade, and union leaders, can contribute an important perspective on managerial attitudes, as well as facts about plant pollution control procedures and the progress of pollution abatement programs.

LEGAL DATA

Most major corporations have been grappling with the problem of meeting new Federal, state and local pollution regulations for several years. While some have achieved, or made great progress toward achieving, the standards, many have been taken to court over their pollution control performance. In 1973, for instance, legal actions were pending or imminent against 40 of the 47 steel mills operated by the nation's seven largest steel companies.

A factory's status with regard to pollution regulations is a major determinant of what will have to be done to reduce pollution. However, legal agencies vary in the amount of pressure they actually exert on a factory to clean up, depending not just on the extent and impact of air and water emissions, but on such hard-to-measure factors as "good faith" shown by the company in the past, the local climate of public opinion, and money and manpower available to the agency. It is thus essential to gather accurate data on both the plant's legal status, and enforcement agency actions.

State Sources

As discussed earlier in this chapter, air pollution programs are primarily a state responsibility. And although the Federal government is presently heavily involved in water programs, the Water Pollution Control Act states that, ". . . it fully intends that the greater portion of enforcement action will be brought by the states." State or local pollution agencies will thus be most closely attuned to a factory's compliance status.

You should begin your research in this area by writing to directors of the state air and water agencies having jurisdiction over the factory you are analyzing, requesting its present compliance status (see list of state water agencies opposite; air agencies are

STATE WATER POLLUTION CONTROL AGENCIES

ALABAMA
Alabama Water Improvement
   Commission
State Office Building
Montgomery, Alabama 36104

ALASKA
Department of Environmental
   Conservation
Pouch O
Juneau, Alaska 99801

ARIZONA
Environmental Health Services
State Department of Health
1624 West Adams Street
Phoenix, Arizona 85007

ARKANSAS
Department of Pollution Control
   and Ecology
1100 Harrington Avenue
Little Rock, Arkansas 72202

CALIFORNIA
California Water Resources Control
   Board
1416 Ninth Street, Room 1140
Sacramento, California 95814

COLORADO
Water Pollution Control Division
Colorado Department of Health
4210 East 11th Avenue
Denver, Colorado 80220

CONNECTICUT
Department of Environmental
   Protection
State Office Building
Hartford, Connecticut 06115

DELAWARE
Division of Environmental Control
Department of Natural Resources and
   Environmental Control
P.O. Box 916
Dover, Delaware 19901

DISTRICT OF COLUMBIA
Water Resources Management
   Administration
Presidential Building
415 12th Street N.W.
Washington, D. C. 20004

FLORIDA
Department of Pollution Control
Tallahassee Bank Building
Suite 300
Tallahassee, Florida 32301

GEORGIA
State Water Quality Control Board
47 Trinity Avenue S.W.
Room 609
Atlanta, Georgia 30334

HAWAII
Hawaii Department of Health
P.O. Box 3378
Honolulu, Hawaii 96801

IDAHO
Environmental Improvement Division
Idaho State Department of Health
State House
Boise, Idaho 83707

ILLINOIS
Illinois Environmental Protection
   Agency
2200 Churchill Road
Springfield, Illinois 62706

INDIANA
Stream Pollution Control Board
   of the State of Indiana
1330 West Michigan Street
Indianapolis, Indiana 46206

IOWA
Water Pollution Division
Iowa State Department of Health
Lucas State Office Building
Des Moines, Iowa 50319

KANSAS
Division of Environmental Health
State Department of Health
Topeka, Kansas 66612

KENTUCKY
Kentucky Water Pollution Control
   Commission
275 East Maine Street
Frankfort, Kentucky 40601

LOUISIANA
Division of Water Pollution Control
P.O. Drawer FC
University Station
Baton Rouge, Louisiana 70803

MAINE
Environmental Improvement
    Commission
State House
Augusta, Maine 04330

MARYLAND
Department of Water Resources
Annapolis, Maryland

MASSACHUSETTS
Division of Water Pollution Control
  Department of Natural Resources
Leverett Saltonstall Building
Government Center
Boston, Mass. 02202

MICHIGAN
Water Resources Commission
Department of Natural Resources
Stevens T. Mason Bldg.
Lansing, Mich. 48926

MINNESOTA
Division of Water Quality
Minnesota Pollution Control Agency
717 Delaware Street S.E.
Minneapolis, Minnesota 55440

MISSISSIPPI
Mississippi Air & Water Pollution
  Control Commission
P.O. Box 827
Jackson, Miss. 39205

MISSOURI
Missouri Water Pollution Board
P.O. Box 154
Jefferson City, Missouri 65101

MONTANA
Department of Health &
    Environmental Sciences
Helena, Montana 59601

NEBRASKA
Department of Environmental Control
Box 94653
State House Station
Lincoln, Nebraska 68509

NEVADA
Nevada State Commission of
    Environmental Protection
201 South Fall Street
Nye Building
Carson City, Nevada 89701

NEW HAMPSHIRE
Water Supply & Pollution Control
    Commission
105 Loudon Road, Prescott Park
Concord, New Hampshire 03301

NEW JERSEY
Department of Environmental
    Protection
Division of Water Resources
P.O. Box 1390
Trenton, New Jersey 08625

NEW MEXICO
Water Quality Control Commission
Environmental Improvement Agency
P.O. Box 2348
Santa Fe, New Mexico 87501

NEW YORK
Department of Environmental
    Conservation
Albany, New York 12201

NORTH CAROLINA
Department of Natural & Economic
    Resources
P.O. Box 9392
Raleigh, North Carolina 27603

NORTH DAKOTA
Division of Water Supply & Pollution
    Control
State Department of Health
Bismarck, North Dakota 58501

OHIO
Water Pollution Control Board
State Department of Health
P.O. Box 118
Columbus, Ohio 43216

OKLAHOMA
Environmental Improvement Agency
Department of Health
3400 North Eastern
Oklahoma City, Oklahoma 73105

OREGON
Department of Environmental Quality
P.O. Box 231
Portland, Oregon 97201

PENNSYLVANIA
Bureau of Water Quality Management
Department of Environmental
  Resources
P.O. Box 2351
Harrisburg, Pennsylvania 17120

RHODE ISLAND
Division of Water Supply &
  Pollution Control
335 State Office Building
Providence, Rhode Island 02903

SOUTH CAROLINA
South Carolina Pollution Control
  Authority
1321 Lady Street, Owen Bldg.
Columbia, South Carolina 29201

SOUTH DAKOTA
South Dakota Committee on Water
  Pollution
State Department of Health
Pierre, South Dakota 57501

TENNESSEE
Tennessee Water Quarity Control
  Board
Cordell Hull Building, Rm. 621
6th Avenue North
Nashville, Tennessee 37219

TEXAS
Texas Water Quality Board
1108 Lavaca Street
Austin, Texas 78701

UTAH
Utah Water Pollution Committee
Utah State Department of Social
  Services
44 Medical Drive
Salt Lake City, Utah 84113

VERMONT
Agency of Environmental Conservation
Department of Water Resources
5 Court Street
Montpelier, Vermont 05602

VIRGINIA
State Water Control Board
P.O. Box 11143
Richmond, Virginia 23230

WASHINGTON
Department of Ecology
Water Pollution Control
P.O. Box 829
Olympia, Washington 98501

WEST VIRGINIA
Division of Water Resources
Department of Natural Resources
1201 Greenbrier Street
Charleston, West Virginia

WISCONSIN
Division of Environmental
  Protection
Wisconsin Department of Natural
  Resources
P.O. Box 450
Madison, Wisconsin 53707

WYOMING
Sanitary Engineering Services
Department of Health & Social
  Services
Division of Health & Medical Services
State Office Building
Cheyenne, Wyoming 82001

listed on page 66).

A factory's position in relation to the law generally falls into one of four categories. You should ask whether it is:

(1) in compliance with existing regulations

(2) out of compliance, but operating under a compliance schedule or "variance" (compliance schedules and variances, under which a company is allowed to continue operations while it upgrades pollution controls, are frequently worked out between a company and a pollution control agency to avoid or settle a lawsuit)

(3) out of compliance, but has no legal action pending against it

(4) out of compliance, and being sued by a government agency.

If you enclose a chart on which factory compliance status can be checked off (see sample), this will help agency officials respond promptly.

You should also request copies of all variances, permits, compliance schedules or lawsuit documents affecting the factory and, if hearings have been held regarding any legal action, of copies of materials submitted in evidence. An agency may refer you to the courts for some legal documents. If it does, be sure to obtain from the agency the docket number and the title of the lawsuit involved, as the courts may otherwise not be able to locate the documents for you.

If the material you obtain from the state agency is incomplete or unclear, or if you wish to know more about any of the legal actions mentioned, you may want to schedule a follow-up interview with an agency engineer or attorney who is familiar with the factory. (If you have already scheduled meetings with other agency officials to go over questions on air or water emissions data, as described in the first part of this chapter, you may want to arrange the interview with enforcement officials for the same day.)

SAMPLE CHART — LEGAL STATUS
(one each for air and water pollution)

PRODUCTION PROCESS	In compliance	Not in compliance No action pending	Not in compliance Lawsuit pending	Not in compliance Operating under variance or compliance schedule

Federal Sources

In cases where state enforcement is inadequate, the Federal government continues to have the right to sue. Between the beginning of 1973 and March, 1974, for example, the EPA initiated 187 enforcement actions under the Clear Air Act, 65 under the Refuse Act of 1899, and 502 under the Water Pollution Control Act.

To find out if the Federal government has taken any action against the factory you are investigating, call the enforcement division at your regional EPA office. You can then learn the details from an attorney in that office. If the enforcement action has resulted in a lawsuit, the attorney will probably advise you to contact the Justice Department lawyer representing the government in the case, who may be able to give you the case documents and discuss the situation with you.

Other Sources

Occasionally, environmental lawsuits have been brought against a company by a party other than a Federal, state or local pollution control agency. Individuals occasionally file suits claiming personal or property damage from pollution. Groups of citizens have banded together and filed "class action" suits claiming that they as a class are being adversely affected by factory emissions. Citizens may file suit to force a company to obey pollution laws, or to require legal agencies to enforce them. If you discover such a lawsuit against the factory you are concerned about, you may want to interview the attorney at the firm handling the case and acquire the relevant legal documents from the firm or the court.

As a final check on the existence of court cases affecting the factory, you should refer again to some of the sources mentioned in Chapter 2. Companies sometimes mention environmental litigation in their annual reports, prospectuses, or press releases. They must describe major lawsuits in their annual SEC 10K forms. In addition, in SEC 8K forms, filed monthly, they must note beginnings and ends of major environmental suits during the month, the parties involved, and the factual basis of any claim or settlement. 10K and 8K reports may be viewed by the public at the Security and Exchange Commission's Washington and regional offices (see chart of office addresses in Chapter 2).

If a visit to the SEC would be inconvenient for you, the *Environmental Law Reporter* published by the Environmental Law Institute, and the *Environmental Reporter,* published by the Bureau of National Affairs, Inc., both try to follow most cases in the environmental field. You may want to refer to one or both publications at a law library.

SPECIAL SOURCES OF ENVIRONMENTAL DATA

The information sources discussed so far in this chapter all contain useful data about nearly every type of industrial operation in the country. However, there are some other governmental and private sources you can tap if the factory you are interested in falls into one of a few special categories, or is conducting certain kinds of activities.

All utilities are required to file detailed reports about their operations with the Federal Power Commission (FPC). One of these reports, "Form 67," contains plant-by-plant air and water pollution control data. This and literally volumes of other information about power plants, such as fuel use, cost of operations, research and development expenditures, and so on, are available for public inspection at the FPC headquarters in Washington, D.C. *How to Challenge Your Local Electric Utility,* a booklet put out by Environmental Action Foundation (Washington, D.C., $1.50), excellently describes special sources of information on power plants.

All aspects of the nuclear power industry, from the mining of uranium ore through the disposal of radioactive wastes, are also regulated by the Federal government. From 1954 until January, 1975, this was the responsibility of the Atomic Energy Commission (AEC). However, the Energy Reorganization Act of 1974 formally abolished the AEC and transferred the agency's regulatory function to the newly established Nuclear Regulatory Commission (NRC), which now has the authority to approve the construction of all nuclear power plants, uranium enrichment plants, uranium mines, and nuclear waste disposal centers. The NRC also monitors operation of these facilities and enforces safety regulations. The agency's offices in Washington, D.C. are thus a good place to go for data on any industrial operation which handles radioactive material.

The Interstate Commerce Commission is another Federal agency which collects information about operations in selected industries. Any corporate action which will cause a change in the amount or type of interstate trade, such as the laying of new rail track

Maine Yankee nuclear power plant. Special data on nuclear generating facilities are available from the Nuclear Regulatory Commission and the Federal Power Commission.

or the construction of electricity transmission lines, must be reported to the Interstate Commerce Commission. Natural gas producers, electric utilities, railroads, and coal gasification companies are all involved in these areas of activity from time to time. The public may review data on these activities at the ICC offices in Washington, D.C.

Many other Federal agencies have a hand in approving and regulating industrial operations. Every time a company wishes to construct a pier, dock, or discharge pipe, or engage in landfill operations which will affect a navigable waterway, coastal wetland, or inland swamp, it must first file for a permit from the U.S. Army Corps of Engineers. The Corps also monitors all industrial dredging and dumping operations. Permit applications can be obtained from regional field offices of the Corps.

The U.S. Geological Survey must approve strip-mining reclamation plans before a company is permitted to extract coal from public land. The U.S. Dept. of Agriculture establishes timber-cutting practices for lumber companies operating on National Forest land. The Bureau of Land Management regulates offshore oil drilling.

The 1969 National Environmental Policy Act specifies that any Federal agency action significantly affecting the environment must be preceded by preparation of an "environmental impact statement." "Action" has been defined to include the kind of Federal regulatory activity described above. (EPA regulatory activity is specifically excluded from coverage, however.) Environmental impact statements, which must include complete data on potential adverse environmental effects of the action and how they can be minimized, may run to several volumes. Locating and obtaining any statements which might contain information about the industrial operation you are studying is discussed in Chapter 6, under "Environmental Impact Statements."

Agencies and commissions associated with the Federal and state governments are constantly studying industrial expansion and its environmental and economic effects. For example, The Northern Great Plains Resource Program, a five-state study group, has issued a series of reports analyzing the potential effects of a massive upsurge in coal mining on the economy, landscape, and culture of the West. The New England River Basin Commission released a report in January, 1975, critiquing industrial development plans affecting Long Island Sound water quality. These reports often contain information about the environmental impact of existing factories or the potential effects of ones proposed. Although it is often difficult to find out about the existence of such reports, as they usually receive minimal publicity, a close watch on the newspapers will help.

Various state agencies other than pollution control departments deal with matters related to industrial pollution and, to widely varying degrees, have involved themselves with environmental problems. Although Fish and Game Commissions in many states limit their role to issuing fishing licenses, Montana's Fish and Game Department is actively crusading for stronger environmental legislation regulating everything from subdivision housing on flood plains to motorboating on public reservoirs. It even publishes a monthly magazine, *Montana Outdoors*, which often discusses the degradation of the state's waters by industrial operations. Maryland also has an unusually active Fish and Game Department. In 1971, it issued a report charging Bethlehem Steel with causing five separate fish kills, involving hundreds of thousands of fish, near its Sparrows Point mill. Bethlehem eventually paid a $16,071 fine and restocked the waters around the mill, on Chesapeake Bay.

Many states, such as California and Colorado, have set up land use planning commissions and commissions to administer state-owned lands. These frequently have jurisdiction over, or require permits for, industrial actions which may affect lake or ocean frontage. State Departments of Natural Resources monitor mining operations in most states from the standpoint of depletion of resources. State Bureaus of Mines frequently enforce safety and environmental protection regulations.

Finally, if the factory you are studying is part of the pulp and paper, iron and steel, or electric utilities industries, three studies prepared under the auspices of the Council on Economic Priorities (CEP), a New York-based public interest research group, may supply you with much valuable information.

Paper Profits (MIT Press, $20), released in 1970, describes the basic technology of pollution control at pulp mills and analyzes, on a mill-by-mill basis, the control systems operating at the 130 pulp mills owned by the 24 largest U.S. pulp and paper companies. This study was updated in 1972 (*Economic Priorities Report*, Vol. 3, No. 3, $3).

The Price of Power (MIT Press, $18.50), presents the same kind of background information and analysis of control equipment and actual emissions as of 1972, for the 124 fossil-fuel and four nuclear power plants operated by 15 U.S. investor-owned electric utilities (including the six largest).

Environmental Steel (Praeger Publishers, $18.50), completed in 1973, similarly analyzes air and water emissions at the 49 iron and steel mills operated by the nation's seven largest steel companies, and describes the possible economic impact of upgrading controls on these firms. CEP, from whom each of these studies can be ordered (84 Fifth Avenue, New York, N.Y. 10011), plans to publish a fourth study in this series, on petroleum refineries, in mid-1975.

NOTES ON RELIABILITY OF DATA

Almost all data on a factory's environmental performance, from whatever source acquired, ultimately comes from the company. While an air pollution control agency official or an OSHA inspector may occasionally come into a plant and take readings of pollution concentrations in stack gases or the workplace, the measurement of pollution by outsiders is the exception rather than the rule.

Company-prepared data exhibits varying degrees of comprehensiveness and accuracy, depending on where it appears. The following generally available data sources are listed in descending order of reliability:

A.) **Materials submitted in evidence to the courts or at hearings on legal actions.**
   Legally required to be correct, mistakes subject to perjury

proceedings. Much time is usually spent in preparing court submissions and they are usually extremely reliable.

B.) SEC 10K and 8K forms and prospectuses for new stock issues and pollution control revenue bonds.
Legally required to be correct and subject to review by SEC personnel, these are sometimes returned to the company for revisions. They are almost invariably filed on time and available as required.

C.) State air agency data and RAPP and NPDES applications.
Legally required to be correct. However the completeness and promptness of filings varies greatly with the vigilance of the supervising agency and the conscientiousness of the company. Some regulatory offices review all materials thoroughly for omissions, inconsistencies, errors, etc., and demand prompt rectification of any problems from the company. Others simply accept materials as filed. Companies vary in the amount of money and staff time they put into maintaining monitoring equipment, making calculations, checking that numbers have been transcribed accurately, etc. Thus, for submissions to contain inconsistencies, mistakes, and omissions, or not to be filed at all, happens frequently enough that one should watch for it.

D.) Published company materials—pamphlets, press releases, etc.
Not subject to government review. Such materials usually do not contain outright factual misrepresentations, but since their purpose is generally to promote the company's image and reputation, they frequently only tell "one side of the story." Omissions of relevant facts and misleading statements or explanations are common.

E.) Newspaper and magazine articles.
Since many news stories constitute second-hand presentations of information drawn from company press releases or interviews, they are even less reliable. The source to which any information in a press article is attributed should be carefully noted, and if possible, consulted to verify the facts.

4

Evaluating the Data

An electric power plant in a medium-sized city is pouring 5,000 pounds of soot and dust into the air an hour, and people living nearby complain of soot on their laundry and windowsills. However, the power company advertises that it has spent $5 million on air pollution controls and is making the best possible effort to reduce emissions. Is it? In another, more rural area, a paint factory is allowing 4 pounds of lead to escape in its waste water into a river each day. Although the river is sparkling clear, local sportsmen claim that the fishing isn't what it used to be. Is this factory discharge harming the river?

Numbers and facts are the best measures of a factory's environmental impact and pollution control efforts, but even the most detailed data is of little help unless it is organized in a way that allows one to draw conclusions about what a factory has done, should do, and could do, about its pollution problems.

This chapter is divided into three sections. The first describes how to arrange the information you have collected into chart form, so that it is easy for you and others to understand. The second explains how to evaluate, using these charts, the present factory record in relation to four performance "yardsticks": legal standards, pollution levels causing adverse health effects, state-of-the-art (best available) controls, and pollution levels in the rest of the industry. The third section discusses how to develop proposals for alleviating factory pollution problems.

ORGANIZING THE DATA

A series of thirteen charts is described below—six on air and six on water pollution, plus one on solid wastes. Each is designed to answer at least one important question about a factory's pollution control performance. The titles of the charts, and the main questions they answer are as follows:

CHART NO.	TITLE	QUESTION
AIR		
1.	Present and Past Air Emissions (lbs./hour)	Are the factory's air pollution emissions increasing or decreasing?
2.	Air Emissions and Legal Standards (lbs./hour)	Is the factory in compliance with air pollution laws?
3.	Air Emissions and Emissions at Other Factories (lbs./ton product)	How do factory air pollution emissions compare to those of similar plants elsewhere?
4.	Design and Tested Efficiency of Air Pollution Controls and Efficiency of the Best Available Controls (% control)	How effective are factory air pollution control systems compared to the best on the market?
5.	Air Pollution Levels In-Plant and Toxic Levels (ppm)	Are factory employees exposed to harmful amounts of air pollution?
6.	Air Emissions and Total Regional Emissions (lbs./hour)	How much of total city, county or regional air pollution does the factory account for?
WATER		
7.	Water Use (million gals./day)	How does the factory's water use compare to the flow of the waterway it discharges into?
8.	Water Pollution Concentrations and Toxic Levels (ppm)	Is the factory discharging harmful levels of water pollution into the waterway?
9.	Present and Past Water Pollution Discharges (net lbs./day)	Are the factory's water pollution discharges increasing or decreasing?

10. Water Discharges and EPA Effluent Guidelines (lbs./ton product)	How do factory water pollution discharges compare to 1977 and 1983 standards?
11. Water Discharges and Present Legal Standards (lbs./ton product or ppm)	Is the factory in compliance with present water pollution laws?
12. Design and Tested Efficiency of Water Pollution Controls and Efficiency of Best Available Controls (% control)	How effective are factory water pollution controls compared to the best on the market?

SOLID WASTE

13. Solid Waste—Amount and Disposal Methods	What are the factory's solid waste problems and how is it dealing with them?

Each of these charts, as you can see, makes a comparison between the factory's pollution record and some measure of performance by which that record can be evaluated. Some charts deal with the factory's legal status, some with its effects on health, and others with the effectiveness of its pollution control systems. Since these comparisons must be made in a variety of ways and units, you will have to do some calculating to put the factory data you have obtained in proper form for charting. However, the required calculations are all limited to basic multiplication, division and percentages, and can thus be undertaken by anyone with a grasp of elementary math and, ideally, the use of a pocket calculator.

The way to construct each of these charts for the factory you are analyzing is explained below. While all of the charts should prove useful, you may, if you are interested in knowing about only one aspect of a factory's performance, such as its legal status, construct only those charts which are relevant to your concern.

To illustrate how the process works for each, sample data from the "Brown Creek Mill," an imaginary factory operated by the imaginary "Staywell Paper Company" has been used to make sample charts. The mill is assumed to produce a type of wood pulp used to make brown paper bags, by a method with a high pollution potential known as the "kraft" process.

Air Pollution

Chart 1: Present and Past Air Emissions. A good way to begin evaluating a factory is to look at recent trends in pollution control. Using the data on present and past emissions you have obtained from your state air pollution control agency, you can construct a chart which will show at a glance where within the factory you are investigating pollution has been getting worse, where it has been reduced, and where it has not changed in several years.

Check your data to make sure it is expressed in pounds of pollutant emitted per hour. This is the unit of measurement used most commonly by air agencies. If another unit is used, check back with the agency on how to convert the data to lbs./hour. The agency should be able to tell you whether the plant operates on a 24-hour-a-day 350-day-a-year basis, in which case you would multiply a tons/year emission figure by 2000 to obtain lbs./year, and then divide by 350 × 24, to yield lbs./hour, or whether the data needs to be converted in some other way.

This and the following four air pollution charts are all organized according to the same basic format. List each factory production process in the left-hand column, and leave a line at the bottom for the factory taken as a whole. In the next column, list the pollution control equipment currently operating for each process. Then complete Chart 1 by making as many additional columns as you have air pollutants. Subdivide each of these columns with dotted lines into two or more (depending on the number of years for which you have collected statistics). You can then insert the appropriate emissions statistics according to process, pollutant, and year, totaling emissions for the whole plant at the bottom.

A sample Chart 1 for the Brown Creek pulp mill is shown opposite. Two pieces of production equipment—the "recovery boiler"

Sample chart; "Brown Creek" pulp mill.

Chart 1

PRESENT AND PAST AIR EMISSIONS

Process	Pollution Control Equipment	Particulates (lbs./hour)		Sulfur Dioxide (lbs./hour)	
		present	1971	present	1971
Recovery Boiler	wet scrubber	125	3775	105	210
Lime Kiln	none	1875	1875	0	0
Power Boilers	none (burn natural gas)	—	—	—	—
Total		2000	5650	105	210

and the "lime kiln" (both used in processing the chemicals which turn wood chips into a pulpy mass)—and the mill's power boilers (used to make steam) are the potentially significant sources of air pollution at the mill. The recovery boiler is equipped with a wet scrubber; however, the lime kiln and power boilers have no pollution control devices operating.

Of the five major air pollutants catalogued by state air agencies, these processes can emit significant amounts of two: particulates and sulfur dioxide. The emissions for the recovery boiler and lime kiln are noted. However, the columns next to the power boilers are left empty. This is because Brown Creek's boilers burn only natural gas, resulting in negligible amounts of particulate and sulfur dioxide pollution. Should Staywell Paper decide, or be compelled because of fuel shortages, to switch to coal or wood wastes as a fuel, emissions from the boilers would increase greatly and have to be noted.

Previous emission inventory data is recorded for the year 1971. Sample Chart 1 shows that Staywell Paper has greatly reduced its emissions from the recovery furnace since that time, probably through use of the wet scrubber, but has made no progress in controlling emissions from the highly polluting kiln.

Chart 2: Air Emissions and Legal Standards. Since legal pressure is often an important tool in bringing about a reduction of a plant's air pollution, it is important to see how a factory's present emissions compare to legally allowable pollution limits. Legal agencies usually set such standards in terms of pounds of pollutant allowable per hour, although such limits are occasionally expressed in other terms, such as parts per million (ppm) in stack gases, or pounds of pollutant allowable per million Btu's (a Btu is a measure of the heat energy produced by burning a fuel).

You can compare emissions from the factory you are studying to the legal limits which apply to it, provided both are expressed in the same unit, in a chart similar to Chart 1. In the two left-hand columns, list production processes and existing controls, and leave columns for each pollutant to the right. Then under each pollutant list both actual and legally allowable emissions for each process.

Sample Chart 2 shows that particulate emissions from the recovery boiler at Brown Creek are in compliance with legal limits, but

Chart 2
AIR EMISSIONS AND LEGAL STANDARDS

Process	Pollution Control Equipment	Particulates		Sulfur Dioxide	
		present	legally allowable	present	legally allowable
Recovery Boiler	wet scrubber	125 lbs./hr	145 lbs./hr	105 lbs./hr	70 lbs./hr
Lime Kiln	none	1875 lbs./hr	180 lbs./hr	0	0 lbs./hr
Power Boilers	none (burn natural gas)	—	.1 lbs./million BTU's	—	1.2 lbs./million BTU's
Total*		2000 lbs./hr	325 lbs./hr*	105 lbs./hr	70 lbs./hr*

*Totals under legal limits are for production processes only, as power boiler limits are expressed in a different, noncomparable, unit.

its sulfur dioxide exceeds standards. The reverse is true of the kiln, where particulate emissions are over ten times those allowable under law.

Chart 3: Air Emissions and Emissions at Other Factories. Big factories often emit more total pollution per hour than small ones, simply because they are larger. Hourly emission figures thus tell you little about how effective pollution control systems are at the factory. To evaluate accurately how well a factory is controlling emissions relative to its size, it is therefore very useful to translate the amount of pollution emitted at a factory per hour into the amount emitted per ton of product produced. This statistic can then be compared to emissions per ton of product elsewhere in the industry.

To obtain the emissions per ton of product of the factory you are interested in, divide pounds of pollutant emitted per hour by the tons of product made in the same time period. (This will probably require taking the plant's annual or daily production tonnage, and converting it to an hourly rate according to the number of hours a day or year the plant operates.)

You can then compare this lbs./ton figure to the maximum amount of pollution the plant could be producing if it used no pollution controls at all. If you have the data, you can also compare it to what other factories in the industry are achieving.

Chart 3 is, again, drawn up in a form similar to Chart 1. However, subdivide the columns for each pollutant with dotted lines into three parts. In the left-hand subdivision under each pollutant note the actual pounds of that pollutant emitted per ton of product produced at the factory. In the middle sub-column, note the *potential* emissions from each process per ton of product. This figure represents the "worst" the factory might be doing, and appears in the EPA *Compilation of Air Pollutant Emission Factors,* described in Chapter 2. Make sure you are using the same units in both columns. Then in the right-hand sub-column, note any data you have been able to find on emissions per ton widely achieved in the industry, or achieved by a particular facility, utilizing a high level of control. The latter statistic should represent close to the "best" level of control the factory could achieve.

A sample Chart 3 for Staywell's Brown Creek pulp mill is shown opposite. This mill produces 1,000 tons of air-dried unbleached

Sample chart, "Brown Creek" pulp mill.   119

Chart 3

AIR EMISSIONS AND EMISSIONS AT OTHER FACTORIES

Process	Pollution Control Equipment	Particulates (lbs./ton of pulp)			Sulfur Dioxide (lbs./ton of pulp)		
		actual	poten-tial*	with widely used controls*	actual	poten-tial*	with widely used controls*
Recovery Boiler	wet scrubber	3	151	47	2.5	5.0	2.5
Lime Kiln	none	45	45	4	0	0	0
Power Boilers	none (burn natural gas)						
Total		48	196	51	2.5	5.0	2.5

*Source: _EPA Compilation of Air Pollutant Emission Factors_

pulp a day, the equivalent, since the plant operates 24 hours a day, of 41 tons an hour. Particulate emissions at the recovery boiler per ton of product are therefore calculated as follows: 125 (the lbs./hour figure taken from Chart 1) ÷ 41 (the tons of pulp/hour) = 3 lbs. of particulates per ton of pulp.

The "potential" emissions are those noted in the EPA *Compilation*. We have also used the *Compilation* to supply emissions per ton of product from control systems "widely used" in the industry. As you can see, Brown Creek's wet scrubber system at the recovery boiler is even more effective than most similar systems used elsewhere. However, the lime kiln, with no pollution controls, is polluting about as much as this process could pollute.

Chart 4: Design and Tested Efficiency of Air Pollution Controls and Efficiency of the Best Available Controls. The effectiveness of air pollution control devices is often measured in another way: in terms of the percentage of pollution in the stack gas which the control device is collecting (and thus preventing from escaping into the environment). To evaluate the effectiveness of the pollution control devices at the factory you are studying, you should compare this "percent removal" figure, known as the system's "operating" or "tested efficiency" both to the maximum efficiency the system was originally designed to achieve, and to the efficiency of the best control systems on the market.

Such comparisons can be made in another chart, similar to Chart 1. The columns for each pollutant are again subdivided with dotted lines into three parts. In the first two sub-divisions, fill in the "tested" and "design" efficiencies of each pollution control device at the factory.

The company or the state air agency should have supplied you with this data. However, if neither one did, or if you want to verify the figures supplied (generally a wise thing to do), you can estimate operating efficiencies using the emissions per ton of product data noted in Chart 3. Divide the factory's actual emissions (left sub-columns) by its potential emissions (middle sub-columns), and multiply by 100, to obtain the percent of factory pollution *not* controlled. Then subtract this figure from 100, for each process and pollutant, to obtain the percent *captured*. You

Chart 4

DESIGN AND TESTED EFFICIENCY OF AIR POLLUTION CONTROLS AND EFFICIENCY OF BEST AVAILABLE CONTROLS

Process	Pollution Control Equipment	Particulates (% control)			Sulfur Dioxide (% control)		
		tested effi-ciency	design effi-ciency	best avail. effi-ciency	tested effi-ciency	design effi-ciency	best avail. effi-ciency
Recovery Boiler	wet scrubber	98	95+	99+	50	50	90+
Lime Kiln	none	0	0	95+	not required	not required	
Power Boilers	none (burns natural gas)	not required			not required		

now have the approximate operating efficiencies of plant air pollution control devices, for each pollutant. They should correspond to any "tested efficiency" figures you have obtained elsewhere.

In the third sub-column under each pollutant in Chart 4, note the efficiency of the best available pollution control system for that process, a figure you should have come across in reading trade publications and other background research materials, or in talking with air pollution agency officials.

Sample Chart 4 for the Brown Creek mill shows that the recovery boiler's wet scrubber is controlling particulates even better than originally expected when installed, and indeed is performing almost as well as the best systems on the market. The lime kiln's pollution controls have an efficiency of 0, since there are none; however, Staywell Paper could install a system capable of reducing particulate emissions by over 95 percent. Finally, power boiler controls are not required, since the boilers are burning natural gas, a "clean" fuel.

Chart 5: Air Pollution Levels In-Plant and Toxic Levels.
In doing your research, you may have uncovered some data on the concentration of toxic pollutants in the air at different places inside the factory. Such levels are often recorded in Occupational Safety and Health Administration (OSHA) inspection reports.

Any ambient in-plant readings you have collected can be organized into another chart, similar in format to Chart 1. However, in the left column where Chart 1 lists production processes, you should list the plant locations where emission readings were taken. Leave the second column blank. On the left-hand side of the column under each pollutant, record the ambient readings. On the right-hand side you can note various kinds of standards or limits, depending on what your research has turned up. If there is an EPA toxic substance standard or an OSHA "threshold limit value" (TLV) for the pollutant, note this. If you have found experimental evidence of levels at which adverse health effects are known to occur, put these figures in this slot, and note your data source in a footnote.

Sample Chart 5 is shown opposite. In the course of our research on Staywell's Brown Creek facility, we have found that the Na-

Sample chart, "Brown Creek" pulp mill.

		Hydrogen Sulfide (ppm)	
Location	Pollution Control Equipment	ambient readings	OSHA TLV
Recovery Boiler		15	10
Lagoon Sludge		5	10

Chart 5

IN-PLANT AIR POLLUTION LEVELS AND TOXIC LEVELS

tional Institute of Occupational Safety and Health (NIOSH) conducted a plant survey, and took readings on hydrogen sulfide levels at various locations. These are recorded, as is the OSHA standard for this gas. Notice that the levels of hydrogen sulfide around the recovery boilers exceed the OSHA limit, indicating that the mill is in violation of the law, and that the health of some workers may be endangered.

Chart 6: Regional Air Pollution Levels. A final important consideration is the extent to which a factory contributes to overall community air pollution. To make a chart showing the contribution of the factory you are investigating, list in the left-hand column the names of pollutants for which you have emission data. In the next column to the right, note total factory emissions (*i.e.* the bottom line figures from Chart 1). In the third column list the amount of each of these pollutants discharged from all air pollution sources in your area, a figure provided by the state air agency or the federal EPA. Again, make sure to use consistent units of measurement.

Dividing the figure in column two by the figure in column three for each pollutant, and multiplying the result by 100, will give you the percentage of total area emissions of various air pollutants accounted for by the factory. Note this percentage in the fourth column.

According to sample Chart 6, the Brown Creek mill accounts for a third of particulate, and over a fifth of sulfur dioxide emissions in the county. While eliminating the mill's pollution might not improve local air quality by a commensurate percentage—since "ambient" pollution levels, *i.e.*, the concentrations in the air, vary with such factors as wind direction and other weather conditions—this chart still indicates the factory's overall contribution relative to other polluters.

Sample chart, "Brown Creek" pulp mill.

Chart 6			
AIR EMISSIONS AND TOTAL REGIONAL EMISSIONS			
Pollutant	Total Factory Emissions (lbs./hr.)	Total County Emissions (lbs./hr.)	Contribution by Factory to County Pollution (%)
Particulates	2,000	6,000	33%
Sulfur Dioxide	105	500	21%

Water Pollution

Like air pollution data, the water pollution data you have collected can also be organized into a series of charts.

Chart 7: Water Use: Discharge Rate and River Flow. To place a factory, whether it is large or small, into the context of the aquatic environment it affects, it is important to know how much waste water the factory discharges and how this compares to the flow or size of the receiving river or lake.

If you have acquired water pollution discharge data from a factory's Refuse Act Permit (RAPP) application, you can find the amount of water discharged from each of its "outfall" pipes in Section II, Item 22, Column 3. Adding these amounts up will give the total factory discharge. If you have obtained factory discharge data from an application filed under the National Pollution Discharge Elimination System (NPDES) program, you will find the amount of water flowing from each outfall pipe listed in Section II, Item 17, Column 3, and the total factory water discharge in Section I, Item 9.

To record this data for your own use, make a four-column chart listing the number of each outfall pipe in the lefthand column, its water discharge in the next column, and either the flow rate of the river, or the capacity of the lake, into which the factory discharges in the third. The latter figure comes from the state office of the U. S. Geological Survey or, if dealing with a power plant, the Federal Power Commission, as explained in Chapter 3. Make sure that the units for columns two and three match; flows are usually expressed in million gallons of water per day. Use the figure for the "low flow" of a river—its flow rate in dry weather.

In the last column, note the percentage of the receiving water body which the water used by each outfall and the factory as a whole represents; these figures are obtained by dividing the flow in the second column by the figure in the third, and multiplying the result by 100.

We can now begin examining the impact of the Staywell Paper kraft pulp mill on Brown Creek. According to its RAPP application, the mill has two "outfall" pipes, which discharge 8 and 6 million gallons of effluent a day respectively. The Creek itself is not large—its low flow at a point slightly above the mill is just 28 million gallons a day. Our sample chart indicates that the mill uses the equivalent of half the Creek's water in its operations.

Sample chart, "Brown Creek" pulp mill.

Chart 7 WATER USE			
Outfall #	Factory Discharge (million gals./day)	Stream Flow (million gals./day)	Percentage of Stream Flow Accounted for by Factory
1	8	28	29%
2	6	28	21%
TOTAL	14	28	50%

Chart 8: Water Pollution Concentrations and Toxic Levels. A factory's waste water may contain few or many pollutants, in large or small amounts. It is very important to know the concentration of various pollutants in factory effluent, and the pollution levels which will cause damage to aquatic life, since concentrations must stay below these levels if a waterway's ecological health and balance are to be maintained.

You can find the concentration at which each pollutant is discharged from each outfall belonging to the factory you are studying in its RAPP application, Section II, Parts A and B, Column 6,* or in its NPDES form, Section II, Item 17, Column 3. All of

---

* Refer also to Section II, Item 22, Column 3 if temperature and pH are potential pollution problems at the factory you are studying.

this pollution may not be caused by the factory in question, however; if the factory is located on a waterway which gets significant industrial use, some or all of this pollution may have been present in the water before it even entered the factory's production processes. The concentration of various pollutants in the *intake* water is noted in a RAPP application, Section II, Parts A and B, Column 1 and in Item 22, Column 1; in an NPDES form, Section II, Item 17, Column 1.

To determine the factory's effect on the concentration of each pollutant in the water it uses, subtract the intake from the discharge concentration. In general, a factory will either increase the concentration, or will have no effect. If the factory has a particularly efficient water pollution control system, however, it may actually reduce the concentration of a pollutant present in the water it uses. In such a case, subtracting the intake from the discharge concentration results in a negative number, indicating the amount of the pollution decrease.

Your data on pollution concentrations can be organized into a multi-column chart which is basically similar in format to the first four air pollution charts described earlier. Begin by listing the outfall numbers in the left-hand column. Next to each one note the production equipment served by that outfall; this should be indicated on a RAPP form, Section II, Item 11, or on an NPDES form, Section II, Item 13. In the third column list the pollution control equipment operating for the outfall, which should be described in the RAPP form Section II, Item 21, or the NPDES, Section II, Item 15.

Now add as many columns on the right as you have different water pollutants to report on, dividing each with dotted lines into three sub-units. In the left-hand sub-division, note the actual intake and discharge concentrations of the pollutant. In the middle sub-division note the change in concentration of the pollutant in parts per million at the factory (discharge minus intake), using plus and minus signs to indicate increases or reductions. Then in the right-hand sub-division, note for comparison purposes a concentration level which is known to have an adverse environmental effect, such as a drinking water standard established by the state or the U.S. Public Health Service; an EPA toxic substance limitation, if one exists; or a concentration level which has been found by scientists to cause adverse biological effects. Check to make sure that any figure you use is also in parts per million.*

A sample Chart 8 is filled in for the Brown Creek mill. Waste water from all processes is mixed together at this mill. The water discharged from Outfall #1 goes through a clarifier providing primary treatment; that which leaves Outfall #2 passes through both a settling basin and an aerated lagoon, providing both primary and secondary treatment.

As noted in the pulp mill *Development Document,* described in Chapter 2, the two major water pollutants at a kraft paper mill are "suspended solids" and "biochemical oxygen demand." (The former is an aggregate measure of solid material leaving the plant; the latter is actually a measure of the capacity of the polluting material in the discharged wastes to use up oxygen in the receiving water during the process of decomposition.)

The RAPP application for Staywell Paper's Brown Creek mill states that the concentration of suspended solids is 50 ppm in the intake, and 65 ppm in the discharge water from both outfalls. The mill thus increases the water's suspended solids content by 15 ppm. Biochemical oxygen demand is 12 ppm in incoming water. In the discharge from Outfall #1 it is up to 212 ppm, a large increase, but not a surprising one since this water receives no secondary treatment. However, the biochemical oxygen demand of the water discharged from Outfall #2 decreases from 12 to 10 ppm. Having passed through the aerated lagoon, the water leaves Outfall #2 "cleaner" of biochemical oxygen demand (BOD) than it was when it entered the plant.

Note that on the sample chart no figures have been inserted under "concentrations causing an adverse environmental effect." This is because such levels cannot be fixed for these two broad measures of pollution. If, however, the plant were discharging several parts per million of a specific chemical like cyanide, one could insert a specific number, in this case a "1," in the column denoting concentration levels causing an adverse environmental impact, since scientific studies have shown that most fish die upon exposure to 1 ppm of cyanide. Fish passing near the

* Except for temperature and pH.

Chart 8

WATER POLLUTION CONCENTRATIONS AND TOXIC LEVELS

Out-fall #	Production Equipment	Pollution Control Equipment	Suspended Solids (ppm)			Biochem. Oxygen Demand (BOD) (ppm)		
			intake/disch. conc.	change in conc. at factory	conc. causing adverse env. effect	intake/disch. conc.	change in conc. at factory	conc. causing adverse env. effect
1	digester, recovery boiler, lime kiln	clarifier	50/65	+15		12/212	+200	
2		settling basin and aerated lagoon	50/65	+15		12/10	-2	

mouth of such an outfall would be exposed to lethal levels, although the cyanide would of course gradually be diluted to
a sub-lethal concentration as it mixed with river water.*

Chart 9: Present and Past Water Pollution Discharges.
In addition to knowing present water pollution levels, you should
be aware of whether factory performance has improved, deteriorated, or remained the same over the past few years. This
can be determined by comparing the number of pounds of
water pollution the factory currently releases to its discharges
in the past.

You can compute the factory's present "net" discharge, this is,
the pounds of pollution it actually puts into its effluent (as opposed to the "total" or gross discharge, which includes both the
factory's contribution and the amount of pollution already in
the water when it entered the plant) from the data already recorded in Charts 7 and 8. The calculations required are not
difficult; however, since a fair number of potentially tedious
multiplication steps are involved, you may want to use a pocket
calculator in this phase of your analysis.

To determine the net discharge of one pollutant from one outfall pipe:

  (1) Take the "change in concentration at your factory" at
      the outfall noted in Chart 8 and multiply by 8.3. (This
      converts the number from parts per million, the equiva-
      lent of milligrams per liter in water, into pounds of
      pollutant per million gallons of water.) Be sure to keep
      any minus sign.

  (2) Multiply the result by the daily flow of the outfall in
      million gallons per day noted in Chart 7.

The result is the net pounds of a pollutant discharged daily from
the outfall. At the Brown Creek mill's Outfall #1, for example,

---

* There are "mixing zone" equations which can tell you the approximate
amount of dilution which will take place at various distances from the
discharge point. However, the calculations can get complicated, and thus
should not be undertaken without the assistance of someone, such as a
water pollution control agency official, who is familiar with them.

Chart 9

PRESENT AND PAST WATER POLLUTION DISCHARGES

Outfall #	Production Equipment	Pollution Control Equipment	Suspended Solids (net lbs./day) present	Suspended Solids (net lbs./day) 1971	BOD (net lbs./day) present	BOD (net lbs./day) 1971
1	digester, recovery boilers, lime kiln	clarifier	996	1,000	13,280	12,500
2		settling basin and aerated lagoon	747	29,000	-100	9300
TOTAL			1753	30,000	13,180	21,800

the suspended solids discharge would be $15 \times 8.3 \times 8 = 996$ lbs./day. At Outfall #2, biochemical oxygen demand is $- 2 \times 8.3 \times 6 = - 100$ lbs./day, with the negative number indicating the net pounds of the pollutant *removed* from the water used daily.

By repeating the above calculation for each pollutant at each outfall pipe, the net discharges from the entire plant can be determined. If you have obtained a RAPP or NPDES application or other material listing factory discharge information for previous years, net discharges for an earlier year can also be computed.

All net discharge data should be recorded on a chart, similar to Chart 8, listing the number of each outfall, the production equipment each serves, and the pollution control equipment operating, in the first three columns on the left. The columns for each pollutant should be divided in half. In the left subdivision list the present net discharge of the pollutant from each outfall. Add up the column and note the total plant's net discharge of the pollutant at the bottom. If you have data on the plant's discharges at some time in the past, note in the right sub-division the discharge of the pollutant at each outfall and for the whole plant for that year.

Our sample Chart 9 shows that the Brown Creek mill effluent discharged from Outfall #1 includes an additional 996 pounds of suspended solid material a day, unchanged for the last three years. However, the amount of biochemical oxygen demand added has increased, suggesting that perhaps the pollution control system has not been well-maintained and its efficiency has deteriorated. The remainder of the mill's waste water, discharged from Outfall #2, shows a dramatic reduction in net biochemical oxygen demand and suspended solids discharged per day in comparison to 1971, indicating that the settling basin and aerated lagoon were put into service sometime in that period. This treatment system now actually discharges effluent which contains a biochemical oxygen demand which is 100 pounds less each day than the water had on entering the plant.

Chart 10: Water Discharges and EPA Effluent Guide-lines: "Best Practicable" and "Best Available" Levels of Control. The Water Pollution Control Act of 1972 required the U.S. EPA to establish two sets of standards for existing industrial polluters: Level I guidelines, to be met by mid-1977, represent the "best practicable" levels of control (BPT) including such factors as economic costs and benefits; more stringent Level II guidelines, to be met by mid-1983, reflect the "best available" controls (BAT) that modern technology allows. The EPA set about determining what these guideline levels should be, and by the end of 1974 had promulgated standards for 33 industries. If the factory you are studying is in one of these industries (see list of *Development Documents for Effluent Limitation Guidelines* in Chapter 2) you should find it extremely valuable to complete Chart 10, and find out where this factory stands in relation to future legal requirements.

When the EPA established "effluent guidelines," it needed some common "yardstick" against which all factories, large and small, could be measured. It therefore decided to set limits in terms of the amount of pollution a factory discharges relative to the amount of product it produces.

A factory's *net* discharge of various pollutants per ton of product is the most accurate measure of a factory's pollution control performance and the EPA has established effluent guidelines on that basis for a few industries, including iron and steel and beet sugar processing. When a net discharge basis is used, this is clearly noted in the *Development Document* for the industry. If the factory you are studying belongs to one of these few industries, you can put your data in a form to be compared to the BAT and BPT standards by dividing the "net pounds per day" figures for each pollutant and outfall recorded in Chart 9, by the factory's production expressed in tons per day. The result is its discharges in net pounds per ton of product.

For most industries, however, the EPA decided that for simplicity's sake it would establish limits in terms of "gross" discharges (the total amount of pollution leaving the factory including what was already in the water when it entered the plant) per unit of product. The EPA felt that using gross figures would be basically fair since at most factories there is not much difference between net and gross discharges, and would greatly reduce monitoring and paperwork problems. If there were a large difference between net and gross (*i.e.*, if a factory had exceptionally dirty intake water), the factory could apply to be evaluated on the basis of its net discharge.

If, like most factories, the one you are studying is regulated on the basis of gross discharges per ton of product, there are two steps you must complete before your data will be in proper form to be compared to the BAT and BPT standards.

First, you must determine the gross discharge (in pounds per day) of each pollutant from each outfall. If you obtained your data from a RAPP application, this information will appear in Section II, Parts A and B, Column 7. However, if you obtained your data from an NPDES application, you must calculate gross discharge using figures noted in Charts 7 and 8. You can also use this calculation to check RAPP data for internal consistency. To calculate the gross discharge of one pollutant at one outfall:

(1) Take the "discharge concentration" of the pollutant at the outfall noted in Chart 8 and multiply by 8.3.

(2) Multiply the result by the daily flow of the outfall in million gallons per day, noted in Chart 7.

Next, you must convert the gross discharge (obtained from the RAPP form or calculated as just described) into gross discharge *per ton of product*. To do this, divide the gross discharge of each pollutant from each outfall by factory production, expressed in tons per day.

At the Brown Creek mill, for example, the gross discharge of suspended solids from Outfall #1 is calculated as follows: $65 \times 8.3 \times 8 = 4316$ lbs./day. Since the mill's production is 1000 tons of pulp a day, the outfall's gross discharge per ton of product would be $4316 \div 1000 = 4.31$ lbs. of suspended solids per ton of pulp.*

---

* In using the result of the computation, (*i.e.*, 4.316), the thousandths place has simply been dropped, rather than rounded off. This is a more "conservative" way to proceed, and a useful rule to follow in doing these calculations.

Chart 10

WATER DISCHARGES AND EPA EFFLUENT GUIDELINES

Outfall #	Production Equipment	Pollution Control Equipment	Suspended Solids (gross lbs./ton of pulp)			BOD (gross lbs./ton of pulp)		
			present	BPT (1977)	BAT (1983)	present	BPT (1977)	BAT (1983)
1	digester, recovery boilers, lime kiln	clarifier	4.31			14.07		
2		settling basin and aerated lagoon	3.23			.49		
TOTAL			7.54	12.0	3.7	14.56	5.6	2.7

Once you have completed your computations of factory water pollution per ton of product for each pollutant and each outfall pipe, these figures can be compared to BPT and BAT standards for the industry. The promulgated standards can be obtained from the issue of the *Federal Register* in which they appear, or from the EPA regional office.

Effluent guidelines in some industries are set up to apply on a process-by-process basis, while in others they apply to entire factories. In the pulp industry, the BPT and BAT standards apply to a factory as a whole, since most pulp mills combine their water wastes from various parts of the plant, treat them centrally, and discharge them at one point. Our sample mill is unusual for the industry, and the pollution loads for each of its outfalls must be added together in order to compare the mill's performance to EPA guidelines.

For other industries which have a wide variety of production processes (*e.g.* steel mills which often have coking, sintering, blast furnaces and steelmaking furnace facilities all at one site) the EPA has set process-by-process effluent guidelines. These specify the pounds of water pollutant permissible per ton of coke, sinter, iron or steel produced at that process. Generally, in such cases, a particular outfall pipe or set of outfall pipes will serve a given process, and this will be noted on the RAPP or NPDES application. If you are dealing with such a facility, you will have to match the appropriate standard to the appropriate discharges. Be sure, when computing the pollution load, to divide pollution by production for the individual process and not the factory as a whole, *i.e.*, divide lbs./day of phenols released at the coke oven, by tons/day of *coke* produced (not tons/day of steel).

A factory's performance can be compared to EPA effluent guidelines on a chart similar in format to Chart 9, with the columns under each pollutant sub-divided into three and labelled "present," "BPT—1977" and "BAT—1983."

According to sample Chart 10, Staywell Paper's effluent at Brown Creek exceeds both the BPT and the BAT standard for biochemical oxygen demand. The company will thus very likely have to add secondary treatment for its wastes at Outfall #1

ards may appear in various units, including parts per million, pounds per day, and pounds per ton of product.)

Sample Chart 11 for the Brown Creek mill shows that the state Department of Environment set standards in gross pounds per ton of product. The mill's suspended solids discharges are safely within current legal limits, but its biochemical oxygen demand is not.

Chart 12: Design and Tested Efficiency of Water Pollution Controls and Efficiency of Best Available Controls.

A final way of pinpointing water pollution problems and analyzing what can be done is to compare the operating efficiency of existing plant pollution control equipment to both the efficiency for which it was designed, and the efficiency of the best control equipment on the market.

Your information on the operating and design efficiency of the factory's water pollution controls may have been obtained from the company or from the water pollution control agency. Your data on the efficiency of the best controls on the market may have come from *Development Documents*, from discussions with a control equipment manufacturer, or from articles in trade publications.

This data should be organized in a chart similar to previous water charts, with three subdivided columns under each pollutant heading. In the first subdivision note operating efficiency of existing controls, in the second column the system's design efficiency, and in the third, the efficiency of the best available controls for that pollutant. "Efficiency" in each case is the percent of the pollutant in the waste water which the system removes.

Sample Chart 12 confirms the high efficiency of Staywell Paper's treatment system at Outfall #2, suggested by the low discharge levels recorded in earlier charts, but indicates that the clarifier at Outfall #1 is not operating at the efficiency for which it was designed. However, even if the clarifier were brought up to its maximum efficiency it would provide only "primary treatment," and thus remove only a small fraction of the biochemical oxygen demand that the best secondary treatment systems could eliminate.

Sample chart, "Brown Creek" pulp mill.

Chart 12

DESIGN AND TESTED EFFICIENCY OF WATER POLLUTION CONTROLS AND EFFICIENCY OF BEST AVAILABLE CONTROLS

Outfall #	Production Equipment	Pollution Control Equipment	Suspended Solids (% control)			Biochem. O. Demand (% control)		
			tested efficiency	design efficiency	best available efficiency	tested efficiency	design efficiency	best available efficiency
1	digester, recovery boilers, lime kiln	clarifier	85	90	90 +	15	20	95
2		settling basin and aerated lagoon	90	90	90 +	95	95	95

Chart 13: Solid Waste—Amount and Disposal Methods.
If you are aware of any solid waste problems at the factory
you are studying, information on the quantity of waste and
method of disposal should be organized into one final three-
column chart. Describe the process which generates the waste in
the left column, the amount and type of waste in the center, and
the disposal system on the right.

To round out our picture of the environmental impact of Staywell Paper at its Brown Creek mill, we have filled in a sample
Chart 13. The mill's water pollution control systems produce
7.5 tons a day of sludge, which the company hauls to a landfill
disposal site.

Sample chart, "Brown Creek" pulp mill.

Chart 13

SOLID WASTES — AMOUNT AND DISPOSAL METHOD

Source	Amount and Type	Disposal Method
clarifier and settling basin	7.5 tons a day of sludge	hauled by truck 10 miles and dumped in a sanitary landfill operated by township

PERFORMANCE YARDSTICKS

Having drawn up your own set of charts organizing factory
air and water pollution data, you are now equipped to come to
significant conclusions about the factory's present pollution control performance. The factory's record can be evaluated in re-

lation to several types of performance "yardsticks". They include:

(1) legal standards

(2) the level of control necessary to prevent adverse effects on local ecological balances or human health

(3) the "state of the art" in pollution control for the industry (*i.e.* best available control technology)

(4) the pollution control performance of other factories in the industry.

You may eventually want to recommend that the factory be brought up to one or more of these levels.

Legal Standards

A major yardstick against which to evaluate pollution levels is legally allowable limits, both present and future. If a company fails to comply with the strictures of the law, it can be taken to court.

To analyze the legal position of the factory you are studying, refer to Charts 2 and 5 for air pollution and Charts 10 and 11 for water pollution. They will tell you whether the factory's air emissions are in compliance with existing emission and occupational health regulations and whether its water discharges meet with present regulations and with standards taking effect in 1977 and 1983.

Refer also to Charts 1 and 9 on present and past air and water discharges to see whether the company has already been upgrading pollution controls (perhaps in anticipation of tighter regulations, or in response to a compliance schedule).

You should now also consider what has been, and is being done about the legal status of each section of the plant. Is the company operating under a compliance schedule? Does it have a water discharge permit? Are any suits already pending? Is legal action imminent? Can compliance be expected soon?

Environmental and Health Effects

Factory emissions can also be assessed in relation to the levels at which adverse environmental or health effects are known to occur. In some cases, these levels correspond exactly to legal limits, but more often than not, adverse effects occur at levels lower than the legal standards.

By reviewing the data contained in Chart 5 on in-plant air pollution levels, and Chart 8, on water pollution concentrations, you can see if there is any direct evidence of pollution which exceeds toxic levels.

If ambient air pollution readings taken at the factory workplace exceed the OSHA or some other safety standard, plant workers are inhaling dangerous quantities of air pollution.

If water pollution concentrations exceed toxic levels, then aquatic life is being damaged around the mouth of the outfall pipe. Although it is more difficult to determine the exact effect of factory water pollution after it mixes with lake or river water, if discharge concentrations are many times the toxic levels and the factory's water use accounts for a substantial portion of the river's flow (see Chart 7), it can be assumed that there will be adverse effects over a much wider area.

High discharges of toxic and "persistent" pollutants, such as lead and other heavy metals and certain organic chemicals, into either the air or water, constitute a particularly serious problem, as discussed in Chapter 2.

You can make some additional judgments about the factory's effects on health and ecology by referring to Chart 6, which indicates the factory's contribution to your region's total air pollution, and Chart 7, on how much of the waterway the factory uses. In your research you might have read papers or reports on city problems caused by pollution. For example, the lung cancer rate may be higher than the national average. If the factory is responsible for 60 percent of the city's air pollution, a rapid improvement in its air control systems may be imperative. If you have found, on the other hand, that the factory is the major discharger into a local waterway and that fishing there is ter-

Two copper mines and smelters have contributed heavily to ambient air pollution east of Phoenix, Arizona.

rific, you might have evidence that its discharges are not adversely affecting the aquatic environment.

The state air and water pollution control agencies may, through their pollution measuring equipment at various locations or through statistical "dispersion models" and "mixing zone" equations, have been able to supply you with additional data on the factory's contribution to ambient air and water pollution levels.

To evaluate the effect on the environment of factory solid wastes, look over Chart 13 to see whether the factory is disposing carefully of by-products, and how it is dealing with any potentially harmful toxic substances in them.

State of the Art

The company you are studying has probably made some effort to control pollution, met with a degree of success, and may be complying with existing laws. However, its pollution control performance might nevertheless be significantly improved if it employed the best available, known as "state-of-the-art," control technology.

Charts 3, 4, 10 and 12 should allow you to evaluate how the factory's pollution controls compare to the "state of the art." Charts 4 and 12 specifically indicate the percent efficiency of existing air and water control systems and those of the best on the market. If Chart 4 shows that a cyclone collector at the factory you are analyzing captures 80 percent of soot and dust emitted from a production process, while electrostatic precipitators on identical processes collect 99 percent, you can safely assume that the factory's system is not "state of the art." Similarly, if Chart 12 indicates that the company's clarifier is removing only 60 percent of suspended solids in factory waste water while pollution control equipment manufacturers have developed clarifiers capable of collecting 90 percent of solids, then clearly the company's clarifier does not provide "state-of-the-art" control. Of course, utilizing the most efficient pollution control devices is not the only way to reduce pollution. Maximum possible pollution control in an industry may involve energy conservation measures or fuel and process changes which result in less pollution being created. Chart 3, "Air Emissions and Emissions at Other Factories," might give you an idea of how your factory is performing in relation to air pollution control at "state-of-the-art" facilities taking all such possible changes into account.

Chart 10, "Water Discharges and EPA Effluent Guidelines" permits an assessment of the factory's overall water pollution performance in relation to "state of the art," since the 1983 "BAT" guidelines noted there are defined as the levels, in pounds of pollutant per ton of product, achievable using the best available controls. The *Development Document*, described in Chapter 2, for the industry explains the various methods, including process changes and pollution control devices, by which BAT performance can be reached.

Achievements at Other Factories

A final yardstick against which to judge a factory's pollution control record, is the record of the rest of the industry. Is the factory you are analyzing one of the dirtiest of its type, one of the cleanest, or just average? Is the company innovative, reactionary, or simply lethargic in controlling pollution compared to competitors?

The performance of the factory you are studying can be compared to the performance of similar plants, in terms of type of pollution control equipment used, the operating efficiency of the equipment, overall emissions per ton of product, the dates when the company began trying to control pollution, its compliance with legal standards and expenditures to prevent environmental harm. Your data on factory performance will appear on your charts, and in the questionnaire on the background of factory pollution control outlined in Chapter 3.

To make accurate comparisons with other factories, you would need the same background data for similar plants in the industry. Lacking the time or opportunity to gather production totals, air pollution inventory statistics, and NPDES or RAPP figures on other factories, you may still make some comparisons using information from sources you have probably already researched.

If the factory under study is a pulp mill, a fossil-fuel electric power plant, or a steel mill, you can refer to one of the studies by the Council on Economic Priorities, mentioned in Chapter 3, for comparative data on industry pollution control measures. If it is in some other industry, you can refer to articles in trade journals, which often describe the type and effectiveness of pollution control equipment operating at various factories. Pollution control manufacturers also sometimes list factories equipped with company-made control systems and describe their operating and design efficiencies. A telephone call to an official at another company or a pollution control agency might provide enough information about another factory's pollution record to compare it with your own.

DEVELOPING POLLUTION CONTROL PROPOSALS

Having thoroughly acquainted yourself with the environmental impact of a factory through your charts on its emissions and discharges, and your research into industry pollution control problems and programs, you are now something of an "expert" on its performance.

In your research, you may have found that the factory you are analyzing is using state-of-the-art pollution controls for all its processes, is having no discernible effect on the environment or human health, and is in compliance with all existing and foreseeable environmental regulations. Such factories exist, and if you are fortunate enough to be studying such a facility you should heartily congratulate the company operating it on its excellent performance. With continued good management and community support and encouragement the factory's record can probably be maintained.

If, however, the factory's performance falls markedly short in one or more of the above areas, you must now evaluate what levels of clean-up are feasible. You should consider the specific programs the company could adopt to improve its pollution

control performance, and analyze their possible economic repercussions. Your conclusions, drawn in as practical and detailed
a fashion as possible, can then be presented to company people,
regulatory officials, and members of the community, and serve
as a basis for working for change.

Establishing Priorities

Your research and charts should have clarified the areas in
which the factory you are analyzing faces its most serious pollution problems. You will no doubt want to concentrate on
encouraging abatement of these first. It is thus often useful to
begin developing proposals by simply listing factory pollution
sources in order of your assessment of their need for control.

The most obvious reason for giving a pollution source a high
priority for abatement is, of course, the sheer size of its emissions. For example, the completely uncontrolled coal-burning
power boilers at a West Virginia steel mill account for nearly
all the over 13,000 pounds of particulates the mill emits hourly;
if you were analyzing this factory, you would probably conclude
that the boilers should be cleaned up first.

Any process emitting a highly toxic air or water pollutant may
also deserve prompt attention, however, since even small quantities of such pollutants pose a real threat to the environment.
Any process whose emissions have a particularly severe effect
on the quality of the workplace environment might also merit
a high priority.

A polluting process located immediately upwind of a residential
area may deserve earlier attention than a process located upwind
of a lake, because its pollution affects a larger number of people.

A problem area might even receive a high priority because it is
easy to control, and the company could eliminate this pollution
quickly and cheaply; or because pollution control equipment is
achieving an operating efficiency way below what is attained
elsewhere in the factory.

Pollution Control Strategies

Once you have ranked the factory's pollution problems according to priorities for control, it is important to consider the kinds of control that can be used for each pollution source.

You will need to refer again to your research into five basic control methods: installing pollution control equipment, changing fuels or raw materials, changing processes, altering operating and maintenance procedures (including alterations to conserve energy), and shutting down a process. As you weigh the feasibility and possible impact of applying each, you may need to return to your original data sources for more information. Don't hesitate to go back and recontact environmental consultants, company officials, regulators, or pollution control equipment manufacturers to discuss timetables and costs of alternatives more specifically.

You will probably find that there are several alternatives for dealing with major factory problems. These constitute the possible "strategies" for controlling pollution at the plant. The ones that finally seem most appropriate to you will depend on whether you want to encourage the company to bring the factory up to legal standards, eliminate health hazards, install best available controls, bring the factory up to par with the rest of the industry, or some combination of the above.

Before assembling a list of pollution control strategies, it is important to make sure that you are not proposing changes that are, for one reason or another, impractical. Before suggesting a switch from high to low-sulfur fuels, for example, be sure that the company will be able to obtain the cleaner fuel. Before recommending installation of a large piece of control equipment, check to see that sufficient space is available in the plant. An ageing plant at a crowded urban location will probably not have room for large holding ponds, for example.

Pay careful attention to the implementation time required for various control options. In some cases, there may be a lag time of several years before a certain control system can be engineered and installed, whereas another type might be operable sooner. The lead time might make the first option impractical.

Similarly, it is generally unrealistic to urge immediate shutdown of an expensive, newly built production process over the installation of control devices, since a company will be loath to retire the process before amortizing its cost.

Finally, before coming to conclusions, you should refer to your background research into factory raw material and energy use, and off-site pollution potential, to consider whether options you would like to propose represent the best possible environmental trade-offs. "Electric" steelmaking furnaces, for example, are intrinsically far less polluting than "basic oxygen" steelmaking furnaces of comparable size. When one considers further that electric furnaces process recycled scrap rather than iron ore, thus helping to alleviate the environmental disruption caused by iron mining and ore processing, they appear to be an even more desirable alternative. However, electric furnaces also require enormous amounts of electricity to operate. If the company has to purchase this electricity from a utility which generates it using uncontrolled coal-burning plants in a highly polluted urban area, then installation of the electric furnace might not represent the best overall environmental trade-off.

In the end, you should have a list of feasible pollution control strategies for each pollution source, several combinations of which, if applied, would result in factory pollution being reduced.

Assessing Economic Impact

The final, crucial, step in preparing a pollution control program is estimating the cost of implementing each control option, since without such estimates it is impossible to project how and when a company might be able to afford improvements.

As noted in Chapter 2, articles in trade journals describing other companies' efforts to control pollution using the same control strategies are good sources of cost data. Environmental consultants and pollution control manufacturers should be able to help estimate the capital costs of pollution control devices. EPA reports contain cost data. The companies which supply alternative fuels or raw materials to other factories in your area are good sources of information on the costs of these materials.

Environmental trade-offs: electric steelmaking furnaces recycle scrap metal into new steel, but use large amounts of power. Above: scrap steel. Below: Electric generating plant, Four Corners, New Mexico.

You should compute the cost of each combination of strategies which, if applied, would eliminate or greatly reduce factory pollution problems. If there are two major pollution sources and two practical ways to eliminate each one, then there are four possible combinations of control strategies, each probably costing a different amount, which the company could adopt.

Taking the combination of strategies with the lowest cost and the combination resulting in the highest cost to the company, provides the approximate capital cost range the company is likely to face in its abatement effort. This information alone can be extremely useful in discussing the feasibility of pollution control with company officials, government people and the public.

However, if time and resources permit, you may want to go on to estimate the "total" economic impact on the company of both the lowest and highest cost combination of clean-up strategies. Its "total" expenses will include operating costs, depreciation, tax savings, interest costs, etc. The company can pass on this total cost through increases in product prices or through reductions in shareholder earnings, or it can reduce profits, or cut costs elsewhere in its operations.

The Council on Economic Priorities model described in Chapter 2 relates the cost of pollution to a specific company's financial future. By plugging your own cost estimates into the equations explained in the CEP study you can arrive at the approximate range of "total" pollution control costs the factory will have to face, and their potential impact on the company's prices or dividends.

Before concluding an economic impact assessment, it is essential to address the issue of whether the company will find it financially worthwhile to invest the necessary funds in controlling factory pollution or whether it may choose instead to close the factory down. This can be a very difficult assignment, as you probably have very little knowledge of overall corporate planning. Nevertheless, you might be able to draw some subjective conclusions about the corporation's intentions with regard to the factory. Since the threat of factory closings is an explosive issue, particularly in times of high unemployment, you must be in a position to assess whether such fears might be justified.

Examining the cost of pollution control equipment needed for a plant in relation to the plant's value is a good first step. For example, at one point in 1973, Kaiser Steel was threatening to close down its enormous Fontana, California, steel works in the face of an EPA clean-up order. However, a close examination of the pollution problems addressed in the EPA order indicated that they could be eliminated for only a few million dollars, and in view of the value of the entire plant, estimated at several hundred million dollars, it seemed unlikely that such costs would necessitate a plant closing. In fact, Kaiser subsequently signed a consent decree in which it agreed to install the pollution controls.

Next, you should examine past company expenditures both on new equipment and on pollution controls at the factory. Recent heavy capital investment in new equipment is strong evidence that the company intends to retain the factory for a number of years, even if additional pollution abatement is necessary.

The product line at the factory could be significant to its future. If, for example, the factory is the only one in the country that makes a product which is in high demand, the company should have no trouble passing along abatement costs through higher prices. It probably will not shut down. The profitability of the factory's products compared to other products made by the company can also be important; a company is unlikely to eliminate a high-profit item from its product line.

The size and quality of the labor pool in your area, convenient access to raw materials, closeness to markets, tax advantages, availability of clean water, room for expansion, and the state of community relations are all factors which might affect your company's decision to invest in pollution controls or abandon the site.

Finally, it is important to keep in mind that, in general, the cost of pollution control is rarely the determining factor in bringing about a plant closing. Age, location, product line, and so on, are usually factors of far greater importance. However, the unavoidable prospect of having to reduce pollution has hastened the closing of certain already marginal plants.

If such a closing seems a strong possibility for the factory you are analyzing, the argument will surely be made (perhaps rightly) that the potential effects on the community of premature loss of the jobs and tax support the factory provides, outweigh the adverse effects of pollution on community health and welfare.

5

Having an Impact

While your analysis of factory pollution control performance, as discussed in Chapter 4, may have shown that the factory you are studying is functioning in a superior manner according to every yardstick of environmental control, it is more likely that the factory failed to measure up in one or more ways. If so, equipped with your charts, with various specific proposals for factory clean-up, and with an assessment of their probable costs, you should now be able to work effectively for improvement in the factory's pollution control performance.

There is, of course, no single procedure to follow to insure that your proposals will be aired and considered. No two factory environmental situations are alike, either in the nature of the pollution problems, the willingness of the company to take action, the vigilance of the regulatory agencies involved, or the degree of community concern for the environment. However, there are certain basic things you can do to get your ideas across.

First, you should discuss your pollution control proposals with officials in the company which owns and operates the factory, since the company must ultimately implement any changes.

Second, since pollution obviously affects the entire community, it is vital that citizens be involved in pollution control decision-making. You should thus present your pollution control recommendations through the press and public forums.

Third, since much of what industry now does in the field of pollution control is determined by government regulation, and government agencies constitute the established institution through which the public can affect corporate behavior, you should present your ideas through government channels.

It is important to take advantage of each opportunity to convey your findings as it occurs, whether it be giving testimony at a water discharge permit hearing, writing an article for a local newspaper, attending a corporate annual meeting, or lecturing at a college or a Rotary Club meeting. However, your efforts should obviously be geared to your resources and capabilities. If you have the advice and help of someone with public relations or journalism experience, you may want to concentrate on presenting your views through the media. If you are working with an organization which includes a lawyer willing to donate legal services, or one which has the financial resources to pay for legal help, you can consider participating heavily in government regulation and lawsuits. If it includes scientists, you can make more technical presentations.

Your specific approach will depend on your analysis of the available options. This chapter describes some of the methods and tactics which have been adopted by citizens and environmental groups across the nation to communicate their views to corporate officials, the public, and regulators, in order to remedy factory pollution problems. These methods should suggest to you publicity and action possibilities which you can utilize.

CORPORATE DIALOGUE

Once you have developed a series of pollution control recommendations, try to arrange meetings with corporate officials, both at the factory and at corporate headquarters. Go first to any company executives with whom you have worked and have a rapport. It is important, in addition, that the meeting include the company's environmental controls director, and one or more company officials in policy-making positions. You should plan to discuss with them the amount of pollution coming from the factory, its environmental impact, your proposals for clean-up, their cost, and the possible impact of these costs on company finances.

Some company officials are anxious to consider local citizens' opinions; many others are not. A few companies have actually undertaken joint planning projects with citizens' groups and regulators, with mutually beneficial results. The management of Amax, Inc., for example, set up a joint committee with representatives of nine different Colorado groups and agencies in 1967 to consider potential environmental problems at a new molybdenum mining and milling operation it wanted to build in the Rocky Mountains near Denver. The committee's work contributed to the company's deciding to move the site for disposal of mine tailings to a location 14 miles from the mine, where contamination of streams and runoff could be controlled; rerouting of access roads and electric power transmission lines to avoid cutting of trees and defacing of the landscape; and employment of certain ecologically sound construction practices. The company, for its part, was relieved at being able to proceed without having to cope with serious legal or media opposition to the project.

With certain other companies, however, persistence and ingenuity may be required to even schedule a meeting. A Chicago citizens' group, Citizen Action Program (CAP), for example, after numerous attempts to open a dialogue with U.S. Steel about a heavily polluting local steel mill, finally announced to the press that it would fight a company vice-president's candidacy for a position in the Presbyterian Church National Assembly, on the grounds that he was avoiding his "moral responsibility for pollution."

The Church Assembly voted to "use its good offices" to set up a meeting between U.S. Steel executives and CAP, and a month later, the President of U.S. Steel flew from Pittsburgh to meet with CAP representatives. At the meeting, the President agreed to allow city inspectors to enter the mill at any time, something which the group regarded as an important concession; and subsequently, under constant pressure from citizens and legal agencies, the company substantially reduced the mill's very high air emissions.

Your degree of success in corporate communication at this point will depend somewhat on the degree of rapport which you established while doing your research. However, even companies which have been totally uncooperative in providing data may want to hear and discuss your conclusions once they recognize the thoroughness of your research and your intention to take the completed analysis to regulators and the public. A corporate meeting will also create a valuable opportunity to verify your data and conclusions one more time, and avoid any risk of error or inaccuracy.

Mitigating Economic Factors

Much of your discussion with company officials may center on the effect of emissions on community health and the natural environment. However, company managers responsible for paying for environmental improvements, and even factory workers, who may lose jobs if costs go too high, may not be persuaded of the wisdom of upgrading controls because of pollution's effects alone. You should thus be prepared to bring up, where relevant, the fact that good pollution control practices may carry economic advantages for a company as well.

Many companies today face expensive and time-consuming environmental lawsuits for failing to clean up air and water pollution, suits which can foster a poor company image among investors and customers. If your charts show that emissions at the factory under discussion presently exceed legal limits, then a lawsuit is a real possibility—a good economic argument for company action. If your charts suggest that emission levels, although in compliance with current laws, may be inadequate to meet future standards, this fact can also argue for a prompt clean-up effort. A company acting in advance of government pressure has more time to consider available control alternatives and install equipment; and in this era of rising labor and raw material costs, a company which begins upgrading now may pay far less for pollution control than one which starts some years hence.

The company may also be able to save money directly through use of pollution control equipment. By cleaning up the in-plant environment, worker illness and absenteeism may be reduced. Plant equipment may deteriorate less rapidly. (U.S. Steel estimates that it incurs about $600,000 worth of damage a year at its Clairton Coke Works near Pittsburgh due to pollution-caused corrosion.) Control equipment which captures valuable raw materials or recyclable by-products may even be revenue producing. Dow Chemical has estimated that pollution control measures at its Midland, Mich. plant caused reductions in raw material leakages

Coke plant: Pittsburgh, Pennsylvania.

and gains in efficiency which were worth $6 million to the company over a three-year period.

A final mitigating economic factor in a major plant clean-up is the public relations mileage such a campaign can generate. If other companies in the industry are being slow to install pollution abatement systems, the company stands to receive considerable favorable publicity for innovative behavior. Voluntary installation of advanced pollution controls can also serve to improve relations between the corporation and pollution control agencies, who might well interpret it as a "good faith" effort.

If other companies in the industry are already acting to reduce pollution, the company will probably want to avoid being publicly known as an industry laggard.

Discussion of all the economic and environmental advantages and disadvantages of the various possible pollution control alternatives will probably take a considerable amount of time. Your meeting (or meetings) will be most productive if you can keep the conversation as specific and factual as possible. Eventually, if disagreements and differences can be ironed out, you should seek a public commitment on the part of the corporation to make specific process changes or to add specific pollution control devices at the factory within a stated time period.

Shareholder Actions

If a meeting with corporate officials proves unproductive or impossible to arrange, you may want to try to communicate with the company's policy-makers through other channels. One such avenue is the "shareholder proposal."

If you own at least one share of company stock, you are entitled to attend the company's annual meeting, vote on all issues presented to stockholders, and raise issues to be put to a vote. Shareholder "actions" have become an increasingly popular way of bringing pressure to bear on a company to change policies, or to disclose information about its operations, particularly in the field of social behavior. Whereas in 1970, approximately 10 shareholder motions aimed at obtaining information on, or changing, company social policies were brought before stockholders, by 1974, there were over 100 such motions. The issues have ranged from placement of "public interest" representatives on the company board of directors to disclosure of data on strip mining practices.

Two organizations can provide extensive information about shareholder actions: the Council on Economic Priorities (New York, San Francisco), and the Investor Responsibility Research Center (Washington, D.C.). The former group publishes several newsletters ($1 each, prepaid) every spring and a more complete report in the fall, all entitled "Minding the Corporate Conscience," which describe all groups and issues involved in making shareholder proposals, and a variety of other citizen "corporate social responsibility" actions as well. The latter group, which provides information primarily to large institutions such as colleges and foundations, publishes a monthly newsletter on shareholder issues entitled *News for Investors* (Washington, D.C., $48 a year).

Not every motion is acceptable for presentation to shareholders. If you decide to submit one, either for disclosure of environmental information or a change in pollution control practices, there is a complex procedural maze (fully described in past issues of CEP's "Minding the Corporate Conscience," $3) through which you must find your way before your motion can reach a vote. As of 1974 just five socially oriented motions had ever been passed by stockholders, all only after receiving management endorsement.

Nevertheless, a shareholder action makes it possible to present your views to the corporation in a public forum and may lead to the opening of dialogue. The prospect of shareholders publicly criticizing the company for environmental neglect, or indifference toward its employees and the people who live near its factories, is something most companies will seek to avoid. As a result, many groups which have filed shareholder motions have found the corporation very willing to discuss issues raised in the proposal, prior to its annual meeting. Amax, for instance, recently sent a representative to meet with members of a church group that had filed a motion asking for information on proposed coal mining on Indian lands.

PUBLIC DEBATE

It is extremely important to have the broadest possible public debate over factory pollution control decisions not only because factory emissions can affect many individuals' health and welfare, but because the trade-offs involved in controlling pollution, such as whether clean-up may cause job losses, can affect and concern the entire community. In addition, a well-informed public is likely to become a concerned monitor of company practices and an important source of pressure for high quality environmental planning.

An important first step in encouraging debate is learning how to work with the press. When you have gathered all your information, had discussions with the company, and have come to a point where you feel you wish to "go public" with your anti-pollution proposals, schedule a press conference to present your findings and recommendations. Write a press release of no more than three pages summarizing your ideas. Be sure to note the time and place of the press conference on it.

The press release should be mailed to all newspapers, magazines, radio and TV stations in the area five days in advance of the conference. If you know the names of specific environmental reporters at these organizations, send materials directly to them. Make follow-up calls to each organization to be sure it received the release, to answer any questions, and to urge that it send a reporter to cover the story. At the press conference, you should make a presentation, not exceeding five or ten minutes, explaining your main points: the extent of pollution, and the type of change needed and possible. The presentation should be followed by a fifteen or twenty minute question-and-answer period.

After a press conference, keep your contacts in the press regularly informed of subsequent developments and progress. If you deliver testimony at a regulatory hearing, send the press a copy with a note of explanation. If an important decision by the company or pollution control agency is made, give the press an analysis of it. For particularly significant events, such as the granting or denial of a variance at the factory you are concerned about, write additional press releases.

Articles in the press will go a long way toward informing people of the primary issues. However, to make detailed information on factory performance and pollution control options readily available, you may also want to write up your research yourself for distribution to community organizations, lecture audiences, or other interested parties. In a 50-or-less-page booklet, it should be possible to discuss the issues and problems as well as your conclusions about factory control alternatives and their costs.

In Pittsburgh, for example, a group called Group Against Smog and Pollution (GASP) prepared a detailed analysis of control alternatives for a highly polluting coke works, entitled the *Clairton Task Force Report* (available from GASP, Pittsburgh, Pa., $3.50), which eventually became the basis for a consent decree settling a government suit against the factory. Government lawyers found it so useful that the state put up the funds to have it reprinted and distributed in the community.

If you are part of a public interest or citizens' membership organization, you may also want to use a newsletter to keep members of your group up to date on current developments and various aspects of pollution problems and their control.

Speaking engagements can be an effective method of communicating your ideas on pollution control at the factory you analyzed as well. You should contact business groups such as the Rotary Club or the Chamber of Commerce, church groups, and political clubs, to offer your services as a speaker at their meetings. You might also be able to prepare a talk that would be appropriate for adult school, college or even high school classes.

Beyond these techniques, you must simply use your own imagination to come up with the best methods of bringing public attention to the issues and fostering debate. GASP, in Pittsburgh, supplemented its public education program with events called "Tours of Pollutionland." The group drove people past the major polluting factories in town explaining the hazards of pollution and what can be done about them. The passengers in the bus were, of course, also presented with the environmental group's views on what could be done to clean up the factories.

In another effort to work with and involve the entire community, the Luzerne-Lackawanna Environmental Council in Wilkes-Barre, Pa. has set up an Environmental Complaint Bureau through which individuals can register complaints about local problems.

CAP, in Chicago, staged a rally at a highly-polluting steel mill in 1972 to bring attention to the company's refusal even to discuss certain pollution controls with the group. About 50 people appeared outside the mill one morning, accompanied by members of the press (who had been notified in advance). While

A GASP "Tour of Pollutionland."

pickets marched around the gate and cameras rolled, the group's leader attempted to gain access to the mill itself to inspect pollution controls and to meet with plant officials. He was turned away by mill guards. The company's uncooperativeness was effectively publicized on the evening news and in the next day's papers.

Some helpful booklets on how to foster effective citizen involvement in a pollution control campaign are *Your Right to Clean Air* (The Conservation Foundation, Washington, D. C., free), *Don't Leave It All to the Experts* (GPO # 0-478-748, $.55), and *Citizens Make the Difference*, by the Citizens' Advisory Committee on Environmental Quality (GPO # 4000-00290 Washington, D.C., $1.75).

GOVERNMENT REGULATION

Government regulation provides an important channel through which citizens can make their views known, and constitutes the official means whereby society establishes the ground rules within which institutions such as corporations function. Your impact on government regulation of the factory you have been investigating can be great, particularly if you have the help of a legal specialist. Citizens are generally involved in three types of government-related activity: hearings, lobbying, and lawsuits.

Hearings

There is a very good chance that you can influence the way the Federal Water Pollution Control Act is implemented at the factory you are concerned about. It is very possible that, with bureaucratic delays, hearings may be held sometime in 1975 to obtain public responses to a proposed NPDES permit worked out between the EPA, the state, and the company. (Although the official deadline was the end of 1974, only 32,383 of over 55,000 applicants had then received their permits.) This permit, described in Chapter 3, establishes the maximum allowable discharge of factory water pollution and usually includes a compliance schedule specifying systematic pollution reductions which must be achieved over a period of years. No legal document affects factory water pollution control more than this permit. Your previous analysis of the permit application, pollution control standards, and water treatment systems puts you in an ideal position to analyze the proposed permit and offer a critique at the hearings.

Two manuals excellently describe how to participate effectively at NPDES hearings: *A Citizen's Guide to Clean Water*, a 50-page booklet published by the Izaak Walton League (Arlington, Va., free) and the very thorough *Water Quality Training Institute Information Kit* prepared by the Conservation Foundation (Washington, D.C. $10), both mentioned in Chapter 2.

If the factory you have analyzed is in the process of negotiating for its NPDES permit, and no public hearing is planned, you may want to petition the EPA or the state agency for a hearing. Hearings must be held whenever there is "significant public interest," and citizens must be notified of the date by legal advertisements in local newspapers at least 30 days before they take place. During this 30-day period, you can obtain a free "fact sheet" from the state or the EPA containing the basic permit requirements. You can review the actual permit itself at either agency.

The proposed permit should be scrutinized thoroughly. Is it completely and properly filled out? How do the proposed maximum effluent discharge levels compare to the actual discharges recorded in the permit application? Do they give the company leeway to increase pollution? Will the compliance schedule result in reduction of discharges to the BPT (Level I) guidelines by 1977? Might the deadlines in the compliance schedule be shortened? What is the environmental impact of the actual and the allowable discharges?

At the hearing, you should present oral testimony and submit written copies of your statements to the hearings board, to the audience, and to the press. Distribution of statements to the press can be particularly effective. A steel mill pollution expert from the Council on Economic Priorities, noting the sparse attendance at the first steel mill NPDES hearing held in the United States, sent copies of his testimony to several newspapers in the county, near Baltimore. Within the next week, several articles reported on the hearings, quoting from his testimony.

Testimony at hearings can vary in substance and content from an emotional appeal for stricter provisions based on community citizen complaints, to a detailed technical discussion of the adequacy of each effluent limitation. In general, the more specific the testimony, the better. In August 1974, a citizen group, called Environment: Pittsburgh, submitted 40 pages of testimony at an NPDES hearing analyzing, sentence by sentence, seven permits for steel mills in the community. The group asserted that effluent limitations were not strict enough to meet guidelines, that monitoring procedures were not sufficient to allow the pollution control agency to gather the evidence necessary for assuring permit compliance, and that compliance schedules were not established in all cases where emissions exceeded effluent guidelines. (Unfortunately, almost a year after this very thorough testimony was delivered, its effectiveness could still not be judged, as Pennsylvania's permit program was still mired in bureaucratic wrangles.)

If the factory has already been granted a permit, there may still be an "adjudicatory" hearing coming up in which you can participate, although this requires the assistance of a lawyer. Adjudicatory hearings, usually requested by the company (but which can also be requested by a citizen group), are a mechanism whereby individuals, the company, or a Federal agency can present evidence appealing the permit's provisions. The regional EPA Administrator then rules on the questions. Over 350 such hearings had been held by the end of 1974, and hundreds more were scheduled for 1975.

You may also be able to have significant input into how other environmental laws (besides the 1972 Water Pollution Control Act) are carried out. If the factory's emissions exceed air pollution regulations, hearings may be held to discuss a compliance schedule for installing air pollution controls. Opportunities and procedures for participating in implementation of the Clean Air Act are thoroughly discussed in *A Citizen's Guide to Clean Air* by the Conservation Foundation (Washington, D.C., free). If the company has applied for a variance from a pollution control regulation for the factory, variance hearings will probably be held. You may want to support denial of the variance, or approval of it only with a strict compliance schedule attached.

If the factory you have analyzed is a power plant, it may be possible to testify about the company's environmental performance at hearings on rate increases. In 1970, local opposition to heavy particulate and sulfur dioxide emissions at Commonwealth Edison's Chicago coal-burning plants led the Illinois Commerce Commission to make a 4.5 percent rate increase contingent on the company's instituting a $200 million clean-up program. The various possible environmental challenges to rate increases are thoroughly discussed in *How To Challenge Your Local Electric Utility*, a booklet put out by the Environmental Action Foundation (Washington, D.C., $1.50).

Your expertise on pollution problems in general may enable you to have an impact not just on the implementation of existing laws, but also on the development of new pollution regulations and legislation (an area in which there has been a great deal of activity in the past five years).

State and Federal hearings may be held to discuss new pollution control bills, proposed effluent guidelines, air pollution "new source" standards, occupational health and safety standards, or upgrading ambient air standards. While you can check on the occurrence of such hearings by checking the local newspaper for posting of required government notices, a more reliable way to keep abreast of hearing schedules is to request being put on the mailing list for EPA and state agency press releases.

At hearings on new laws and regulations, you might be able to offer evidence of adverse environmental effects that need to be prevented, poor plant emission control that needs to be regulated, or ineffectual pollution control agency efforts that could be improved through stronger legislative directives. Talk to the public officials that represent you on the local, state, and Federal level to see where you can have input into the legislative process.

Lobbying

If bills under consideration seem inadequate, you may want to lobby for stronger legislation. Political Action for Clean Air, a New York City activist group, meets regularly with local Congressmen to clarify its positions on pending legislation. It also endorses and actively campaigns for candidates who take strong environmental stands. A helpful booklet on environmental lobbying techniques is the Sierra Club *Political Handbook* (San Francisco, $.50). You might want to discuss how to lobby effectively with some of the groups who are involved in such activities. Environmental Action and the Environmental Policy Center, both in Washington, D.C. coordinate Federal environmental lobbying efforts.

Lawsuits

In addition to participating in hearings and lobbying activity as described above, you can use the courts to effect change.

It is possible for citizens to file suit either against a pollution control agency for not enforcing the law, or against a company for not obeying it or for damage done by pollution to property or life. Legal actions require expert legal advice and are normally very expensive and time-consuming undertakings. Nevertheless, hundreds of important environmental cases have been brought and won by environmental groups since 1970.

Two nationally prominent public interest law firms—the Natural Resources Defense Council (NRDC) with offices in New York City, Washington, D.C., and Palo Alto, Calif., and the Environmental Defense Fund (EDF), located in East Setauket, Long Island, N.Y., Washington, D.C., Denver, Colo., and Berkeley, Calif.—have been extremely active in bringing environmental lawsuits against the government and polluters under air and water pollution control laws. They have achieved a variety of significant results, from banning of DDT to faster issuance of water pollution regulations. You can contact these groups, or another law firm which handles environmental litigation, to discuss whether there have been any violations of the law at the factory you are concerned about which might serve as grounds for a lawsuit.

Both the Clean Air Act and the Water Pollution Control Act grant the public the right to sue the government or a company in Federal court if they miss deadlines or fail to fulfill their legal responsibilities. The court can then compel agency or company compliance. Nine states (New Jersey, Connecticut, Florida, Indiana, South Dakota, Massachusetts, Michigan, Minnesota, and California) also have laws which allow citizen lawsuits against environmental abuse.

A good reference book on citizen environment lawsuits is *Environmental Law* by Arnold Reitze, Jr., available in many libraries. You might also want to refer to an excellent pamphlet, *Citizen Suits Under the Clean Air Act*, put out by NRDC (New York, N. Y., free).

An individual can sue a company for pollution-caused damages via either a civil damage suit or a class action suit. In a civil damage suit, you would claim to have incurred property damage or personal harm because of factory pollution. A Christmas tree grower in Virginia, for example, sued the local power company on the grounds that the sulfur dioxide emissions from one of its generating plants was stunting the growth of his trees. Unfortunately this type of suit generally does not exert much pressure on a company to clean up since paying for damages usually turns out to be much cheaper than abating pollution. However, if the case should receive a lot of publicity, a company might choose cleaning up over tarnishing of its public image.

A class action suit can bring more substantial pressure against a company. Such suits can be brought by one or more individuals claiming to represent a large body or "class" of people who are in a similar position. For example, one person in a neighborhood may be able to file a class action suit against a company on behalf of everyone in the neighborhood, on the grounds that all have been adversely affected by factory pollution. The value of a class action is that a successful suit forces the company to pay damages to everyone in the class. The cost to the company of losing the case is thus many times greater than the cost involved in paying off a conventional civil suit. Class action suits can also call for a court injunction against continuance of pollution.

Before going to court for any reason, you should consult a law firm that handles environmental cases and might be sympathetic to your position and financial limitations. The Council of New York Law Associates in New York City specifically arranges for volunteer legal assistance to non-profit groups in the New York–New Jersey–Connecticut metropolitan area, for *pro bono* projects. In other areas, the local Bar Association may be able to provide similar help.

Whatever the techniques you decide to utilize, it is important to be aware from the start that the process of raising environmental consciousness is often slow, and the process of real change is even slower. Occasionally, radical improvements in environmental quality can be accomplished within a few years. All too often, however, progress comes in small increments, with an environmental group's vigilance necessary at every step.

Nevertheless, there have been remarkable accomplishments in overcoming severe factory pollution problems. In Missoula, Montana, for instance, a combination of efforts, beginning in 1967 with picketing of the highly polluting kraft pulp and paper mill there, and including an Environmental Defense Fund lawsuit, a strengthening of state air regulations, and a two-page color spread in *Life* magazine ("residents call the mill 'Little Hiroshima'") culminated in 1970 in a $14 million commitment by the Hoerner Waldorf Company to improved air pollution controls. By 1972, there were significant improvements in air quality in the Missoula Valley. If there is a moral to this, it is that for those who evince a sustained concern for a clean environment, the mechanism for enormous change exists.

6

New Plants: Preventive Medicine

New factories, like old ones, have the potential to cause air and water pollution. In addition, the building of a new plant may raise environmental issues not encountered when considering an existing factory's impact. A new factory requires land, which may presently be "open space". It will consume energy and raw materials and use water, which may presently be unused or allocated elsewhere. It will create new traffic patterns which can affect air pollution levels. It will provide new jobs, which can radically alter the social environment of a community. And it may encourage the location of related plants nearby, magnifying all these effects.

Factories being built today often carry a relatively large potential for affecting the environment. In many industries, companies are now building new facilities in giant sizes. Moreover, many new plants use new technologies, which can present unrecognized or unresolved problems of pollution control and public safety.

But while new plants sometimes threaten to be unusually serious polluters, new plants also present an unusual opportunity to exercise sound environmental planning. Pollution control systems are generally less expensive, more effective, and easier to install if built with the plant at the start than if added on after completion. Prior to construction, it is possible to select as the location for the plant a site that is suitable for industrial development. Companies are free to choose production processes and product lines taking into account their pollution potentials. Environmental impact studies can be conducted which can clarify the scope of potential pollution problems and be used to develop plans to minimize factory pollution.

New metal casting plant at Flat Rock, Michigan, operating at full blast. Control equipment amounted to 10 percent of plant cost.

In the future, it should be increasingly possible to prevent many factory pollution problems from ever developing, by employing sound environmental planning in the design and construction of new plants and additions to old plants. You should thus learn as much as you can, as early as possible, about a new factory in order to be able to influence plans while they are still being formulated.

Official applications for permits to construct a new factory may cost hundreds of thousands of dollars to prepare and may be based on several million dollars' worth of research and planning. By the time applications have been completed and submitted to government agencies, a company may have made a strong commitment—both internally and publicly—to building the proposed plant substantially as described in the applications. Or, a company may take the position that modifications of or alternatives to the proposed plant, while perhaps desirable in theory, cannot be researched and constructed in time to meet the need for the products to be manufactured. Many electric utilities and gas companies, which produce essential goods for which demand has been increasing in recent decades, have made this argument.

This chapter discusses the special sources of information, such as "Environmental Impact Statements," available for new plants, as well as the special areas, such as land use, construction, energy use, and social impact, in which a new plant's potential effects should be assessed. The chapter also deals with potential impact on air and water quality. As the text suggests, many of the information sources and methods of analysis discussed in Chapters 2 through 4 are applicable to new plants in these areas. The material presented in this chapter should be used in conjunction with the material presented elsewhere in this book to a greater or lesser extent, depending on the new plant's stage of development, and the kinds of data that are available on it.

BACKGROUND INFORMATION

Before you can assess the potential environmental impact of a proposed industrial plant you must first, of course, know of the proposal's existence. While few industrial projects are advertised in the early planning stages, there are indicators or "signals" of potential industrial development which you can look for. County land transaction records, for example, may indicate that a company is buying up property adjacent to an existing plant, possibly for expansion purposes, or that a company is assembling parcels of vacant land upon which to construct a new facility. (In some cases—generally in an effort to avoid land speculation— land purchases may be carried out by "dummy companies" or real estate brokers quietly acting on a company's behalf.) The minutes of local zoning meetings and hearings may indicate that a company is seeking zoning changes to enable it to build a factory.

The Nuclear Regulatory Commission (created out of the split-up of the AEC) requires utilities planning to build a nuclear power plant to operate a meteorological monitoring station for one year at the potential plant site prior to filing an application for a construction permit. Many state utility commissions also now require electric utilities to file information on their long-range (10-year or more) construction plans, including an inventory of potential plant sites.

Once a company has made some concrete plans to build a factory, information becomes easier to obtain. Such plans are often reported in local newspapers or in trade journals. Local government officials may be aware of some of the details. The company itself may describe and publicize the project in press releases, public relations materials and its annual report, available free by writing or calling the company.

In looking over all publicly available materials, you should keep certain questions in mind. First, what kind of product will the proposed factory manufacture? If it will be a light industrial factory, such as an electronics plant, or a facility which assembles products from components produced elsewhere, then it is unlikely to have a major environmental impact. If, however, it will be a heavy industrial plant, such as a paper mill, chemical plant or an electrical generating station, then the pollution problem may be significant. If you can, find out the Standard Industrial Classification number applicable to the facility (as described under "Industry and Factory Production" in Chapter 2).

Next, what is the factory's proposed size, in terms of units per day of production? Generally, the larger the facility, the greater its pollution potential.

What raw materials will it use? A coal-fired plant can be expected to have a more severe environmental impact than one fueled by oil or gas. A pulp and paper mill using virgin lumber may require logging of local forests; a pulp mill which recycles waste paper avoids such an impact, although it may create more serious water pollution problems.

How many and what kinds of jobs will be made available by the facility, both during construction and in actual operation? New plants creating large numbers of new jobs for local workers can be a major economic asset to the community which outweighs environmental liabilities. However, if the new plant will require the migration of a large work force to your area, this can create a new set of social and environmental problems with which the community must contend.

Once you have answered some basic questions about the proposed factory, you should try to obtain a good background understanding of the company and industry, how the factory's product is produced, environmental problems which may result, methods of controlling such problems, and what such controls may cost. Information in these areas can be gleaned from a variety of sources, fully discussed in Chapter 2.

A good place to begin is with the company which will operate the factory. Is it privately or shareholder-owned, the subsidiary of a larger company, or perhaps government-owned? Is the company large or small in terms of the industry of which it is a part? Is it highly profitable or barely surviving financially? Is it nationally or locally based? What are its pollution control policies? Such questions can be answered through a combination of sources discussed in Chapter 2, including the company's annual report (if it is an investor-owned corporation), articles in business periodicals, prospectuses prepared in conjunction with sales of bonds or stock to finance the proposed facility, and directories prepared by the industry's trade association.

Next, you should study the methods by which the industry to which the proposed plant belongs makes its product, in order to understand the types of pollution problems which may occur, and available methods for controlling them. Descriptions of industry technologies are available from industry trade associations and other references described in Chapter 2. Such descriptions also

Barge near Georgetown, South Carolina. Pulp mills may require logging of local forests.

appear in two types of U.S. Environmental Protection Agency (EPA) documents which, if they have been prepared for the industry to which the proposed plant belongs, can be extremely useful to you at every stage of your analysis. These are the *Background Information—New Source Performance Standards* reports, which EPA has issued for 18 industries (see list, p. 51) in the process of setting air pollution standards for new plants; and the *Development Documents For Effluent Limitations Guidelines and New Source Performance Standards* prepared by EPA for 40 industrial categories (see list p. 14) in the process of setting water pollution standards.

While learning about the company and industry, you should give attention to plants operating elsewhere which are similar to the proposed factory. They can provide an indication of the types of pollution problems which the proposed plant may create, as well as existing pollution control methods which may be applicable. Such plants may be mentioned in the company's annual report, annual reports of the industry trade association, in trade publications which cover the industry, and industry directories (all described in Chapter 2).

If the proposed plant turns out to be the first of its kind, you may want to insist that the company or an appropriate government agency conduct a particularly rigorous assessment of potential environmental problems in the early planning stages, when corrective action is easier, since experience with the technology is not available.

By this time, you will probably be aware of pollutants which the proposed factory may emit. You can learn about at what levels and in what ways they can affect the environment by consulting the sources described in Chapter 2 under "Effects of Pollution." Be sure to consider all pollutants, including both those for which some legal standards have been established, such as the "criteria" air pollutants and the water pollutants mentioned in the effluent guidelines, and other toxic pollutants for which no specific standards exist.

Finally, you should have some idea of the possible costs of pollution control for the proposed factory. While costs of controlling pollution generally run much lower at a new plant than at an existing one, since controls can be included in its basic design, their cost can nevertheless amount to many millions of dollars. The EPA has estimated, for example, that it will add $5.5 million to the cost of a new zinc smelter to meet Federal air standards, the equivalent of 7 to 11 percent of the plant's total cost. Although the costs of certain kinds of sound environmental practice discussed in this chapter, such as land use planning, or "good housekeeping" at a construction site, are sometimes minimal and often difficult to quantify, it is nevertheless important to keep costs in mind. They will certainly affect company thinking about particular pollution control proposals, and even whether the company will locate at a particular site at all. The best sources of data on the costs of air and water control alternatives at new plants are the EPA *Background Information* and *Development Documents* reports. Additional sources are discussed in the section of Chapter 2 on "Economics of Pollution Control."

ENVIRONMENTAL IMPACT STATEMENTS

Having acquired a general understanding of both the company and industry to which the proposed plant belongs, you can begin gathering specific data to analyze the factory's potential environmental impact and how such problems can be minimized. An excellent source of data, although one available in all too few cases, is an "environmental impact statement." If the proposed factory requires a Federal permit or will be built in the state of California, an "environmental impact statement" (EIS) will probably already have been prepared, or at least be in the process of development.

Federal EIS's

The landmark 1969 National Environmental Policy Act (NEPA), specified that a Federal agency must publish an environmental impact statement before taking any action "significantly affecting the quality of the human environment." NEPA gave to the President's Council on Environmental Quality (CEQ) the job of reviewing such statements. The result is that an EIS must now be filed for any project which is Federally sponsored or which requires a Federal permit (with the exception of EPA regulatory permits and other EPA regulatory activities, which the courts have deemed provide the functional equivalent of an EIS); and an EIS must be filed for any state, local or private project receiving significant Federal funding. A number of states have since adopted "little NEPAs" which call for the same type of analysis of state actions. Around 1,300 EIS's are now prepared annually under the Federal NEPA. To date the major filers have been the U.S. Army Corps of Engineers, mostly for its many dam-building and stream channelization projects, and the Department of Transportation for highway projects.

However, certain categories of industrial plants, particularly those in the energy field, must also be the subject of Federally-prepared environmental impact statements. They include new nuclear power plants, licensed by the Nuclear Regulatory Commission; new hydroelectric and pumped-storage generating plants, licensed by the Federal Power Commission; and new liquefied natural gas terminals, also licensed by the FPC. In addition, any company planning to build a factory in a coastal wetland or inland swamp must now receive a permit from, and thus be the subject of an EIS by, the U.S. Army Corps of Engineers.

Even where only a part of a project requires a Federal license or permit, an EIS may still be prepared. For example, a proposed paper mill whose construction involves building a pier or dredging a channel will require an EIS by the Army Corps of Engineers, although a mill which would not alter a waterway probably would not.

The CEQ has specified that an EIS must contain, among other things, sufficient technical data so that other parties can make an independent assessment of environmental impact, a description of all unavoidable adverse impacts, and an exploration of alternatives to the action. In practice, this means that an EIS for an industrial plant will generally present a detailed description of production processes at the proposed facility, of projected raw material and water use and product output, of planned air and water treatment systems, and of projected pollution impact. Most factory EIS's cover anticipated levels of air and water pollution, amount of solid waste to be generated, noise levels, and land use effects, such as impact on archeological sites or on habitats for endangered species. An EIS for a pilot coal liquefication plant in West Virginia prepared by the Office of Coal Research of the U.S. Dept. of the Interior, for example, explains how the plant will make fuel oil from coal, and states that the plant will use 20 tons a day of high-sulfur eastern coal, 450,000 cubic feet a day of purchased natural gas, and 10,000 to 20,000 gallons a day of water supplied by a well. It further indicates that the plant will produce 9.2 tons a day of desulfurized fuel oil and a variety of byproducts (also quantified), and lists the parts per million of hydrogen sulfide, sulfur dioxide, nitrogen oxides, and carbon monoxide to be expected in the gases it will emit into the air, and of phenols, organics, cyanide, and salts in the water it will discharge into the Ohio River.

Although neither NEPA nor the CEQ specify exactly when an EIS must be published, the CEQ guidelines require that it be developed early enough "to permit meaningful consideration of the environmental issues." The guidelines also require: that no final action be taken on the proposal until 90 days after an initial or "draft" EIS is filed; that comments from related Federal agencies, state and local governments, and interested private organizations including businesses and environmental groups be reviewed in preparing a final EIS; and that at least 30 days elapse after filing of a final statement with the CEQ before the agency acts on the proposal. This insures a three-to-four month period at the very minimum between publication of an initial draft EIS and breaking of ground.

A Federal EIS can obviously be an extremely valuable source of data. If available, you should obtain it for the proposed factory you are studying. The CEQ must notify the public in the *Federal Register* of publication of a draft EIS. The agency preparing it may also provide notice of the availability of a draft EIS in local newspapers and may hold public hearings to elicit comments.

If you are unsure, however, as to whether there is a Federal EIS for the factory you are interested in, you can contact the company, the CEQ NEPA coordinator (722 Jackson Place, Washington, D.C. 20006) or a local law firm which handles environmental cases. Individuals at one of these sources should know whether an EIS is required, published, or in preparation. CEQ also publishes the *102 Monitor* (named after Section 102 of NEPA which contained the EIS requirement), at the address above, which provides a monthly list of projects for which EIS's have been prepared, available at $10 a year.

Attorneys can also, if necessary, give you advice on the possibility of legal action to compel an agency to file an impact statement. Citizens groups have gone to court in numerous instances to require agency preparation of comprehensive statements. When the U.S. Dept. of the Interior published an EIS for the Trans-Alaska Pipeline which omitted detailed discussion of much of the Pipeline's environmental impact, for example, the Wilderness Society and others sued to force the Department to prepare a complete EIS. The result was a nine-volume statement which took over a year to produce. While the delay may have contributed to cost increases in building the pipeline, it also led to significant design modifications and rerouting, to deal with freezing, thawing and possible earthquake problems not disclosed in the first EIS. A good booklet on NEPA and how its legal requirements are being interpreted is *The National Environmental Policy Act in the Courts*, by Prof. Harold Green of George Washington University National Law Center (available from the Conservation Foundation, Washington, D.C., $1.00 prepaid).

The environmental impact statement itself can generally be obtained from either the company sponsoring the project or the Washington or regional office of the Federal agency which prepared it. Most agencies will provide a free draft EIS to anyone wishing to comment. Where fees are charged, they usually run about $5 per 300-page volume. If the impact statement was prepared over a year ago, the agency may refer you to the National Technical Information Service (5285 Port Royal Rd., Springfield, Va.) for a copy. As a last resort, the private Environmental Law Institute in Washington, D.C. will copy hard-to-get EIS's for interested groups at $.10 a page.

State EIS's

Following the Federal government's lead, twenty-one states as of early 1975 also had legislation or administrative orders requiring preparation of impact statements in certain circumstances (see list of states). The most notable of these is California's amended 1970 Environmental Quality Act, which requires an environmental impact report (EIR) for all projects, whether carried out by state government, local government, or a private entity, which will have a "significant effect on the environment." This very broad application of the EIS concept to virtually any activity in the state has resulted in approximately 6,000 EIR's being filed annually, and means that one should be prepared for any factory to be built in California. A "notice of completion" of an EIR should be (but isn't always) filed with the state Resources Agency in Sacramento, and the statement itself should be submitted to the relevant local agency.

STATES REQUIRING ENVIRONMENTAL IMPACT STATEMENTS

States with Comprehensive Statutory Requirements

California	Montana
Connecticut	North Carolina
Hawaii	South Dakota
Indiana	Virginia
Maryland	Washington
Massachusetts	Wisconsin
Minnesota	

States with Comprehensive Administrative Orders

Michigan
New Jersey
Texas

States with Special or Limited Requirements

Arizona (water-oriented projects)
Delaware (projects requiring permits under its coastal and wetlands laws)
Georgia (toll road projects)
Nebraska (highway projects)
Nevada (electric power plants)

In the other 20 states with EIS requirements it is less likely that an EIS will have been prepared for a proposed factory. Only Massachusetts, Washington, Michigan, Montana and Wisconsin specifically demand a statement on private activities requiring a state permit. No more than a few hundred impact statements are prepared annually in each of these states. The addresses of the state agencies which review, or can answer questions about, state EIS's are listed in the CEQ's fifth annual report, *Environmental Quality—1974* (GPO #4000-00327, $5.20). This book also contains a good overview of Federal and state NEPA programs.

Company EIS's

Even if not specifically required to by law, many companies will prepare an assessment of the potential environmental impact of a proposed factory for their own internal planning purposes, or would be willing to work with local concerned citizens to develop one. Particularly when building a new plant, a company will generally be anxious to get off on the "right foot" in the community, and work out any environmental problems before construction begins.

An internal company assessment will almost certainly cover air and water pollution and pollution control cost factors, and may explore land use, energy, solid waste, construction and social impact as well.

To learn about what kind of environmental assessment the company has conducted to date, write to the president of the company or the head of the division which is planning the facility. Explain that you wish to discuss environmental questions with them and describe the areas you wish to cover.

If you can arrange an interview, it is best to come equipped with as many specific questions as possible. In the areas of air and water pollution, you should request all the information described in Chapter 3, including data that would normally be filed in an air "emissions inventory," an NPDES permit application, and the data asked for in the sample company questionnaire. You may also want to ask for data on potential problems (described in later sections of this chapter) related to land use, construction, energy use, and social impact.

If in an interview you develop a good rapport with company officials, and illustrate your willingness to proceed with an open mind and an awareness of cost factors to seek the best possible resolution of environmental issues, you may well be able to establish a working relationship which will allow you to have input into company decisions and actions throughout the planning and construction process.

LAND USE

The first area you should consider in evaluating a new factory's environmental impact is the appropriateness of the choice and projected use of the site under development. If the proposal for the plant is still in the early planning stages, then the question of siting may still be open. If plans are more advanced and a commitment to a site has been made, then the projected utilization of the various portions of the site should be evaluated.

There are only a few state—and no Federal—agencies responsible for insuring environmentally-sound siting of industrial plants (see list of agencies). Land use is therefore an area in which citizen participation can be particularly important.

To begin your research, you will need to obtain a map of the proposed site which shows the location of the factory facilities, man-made features such as rail lines, and natural features of the surrounding area. Such a map may have been prepared by a governmental agency for an environmental impact statement, or by the company for its own use or for use by a local zoning board, planning authority, or land use commission. If a site plan is unavailable from any of these sources, you can obtain a topographic map of the area from the U.S. Geological Survey. (Write to the U.S.G.S., Federal Center, Denver, Colorado requesting the free index to topographic maps for your state. You can then use this to order the map you want for $.75.)

You should try to determine the site's present owner (if it is not already company-owned). Although probably in private hands, some land may be held by a government agency such as the Forest Service or Bureau of Land Management, in which case the public may have more say through hearings or other legal channels in how it is used.

Try also to learn from the company the specific intended uses of each part of the site, such as for boilers, ovens, storage tanks, solid waste disposal, roads, parking, or buffer zones.

A detailed analysis of whether the company's choice of, and possible utilization of, the site represent the best possible decisions in terms of your area's physical and social environment is some-

---

STATES HAVING INDUSTRIAL SITING REVIEW BOARDS

General Siting Review

Vermont   Regional review of new industrial (and other) developments of substantial size.

Industrial Siting Review

Maine   State Environmental Improvement Comm. reviews new industrial development.

Power Plant Siting Review

Arizona	Montana
Arkansas	New Hampshire
California	New York
Kentucky	Ohio
Maryland	Oregon
Massachusetts	Virginia
Minnesota	Washington

Coastal Zone Siting Review

California   Statewide commission and regional panels review new industrial (and other) developments on coast.

Delaware   New heavy industry banned within 2 miles of coast. Other industry must obtain permit for coastal development.

thing best undertaken by trained planners and scientists. Ultimately you may want to propose that a government agency or the company undertake such a study. However, there are certain obvious factors which a layperson can and should initially consider.

Will the proposed plant be located near a major population center? If so, you should give particular attention to local air pollution effects, and any potential for a catastrophic accident such as explosion, fire, or release of toxic materials. Does the land include a particularly valuable ecological resource, such as a tidal estuary (breeding ground for much ocean life) or a habitat for endangered species? Or does the land have particular historic value, containing an archeological site for example? In that case, you should determine whether the proposed factory will significantly alter or damage the resource and what changes would be necessary to prevent this. Is the land particularly subject to any natural hazards, such as hurricanes or earthquakes, or is it part of a natural flood plain? If it is, unless precautions can be taken, alternative sites should be considered. Is the land important for its renewable resources, such as watersheds, aquifer recharge areas (underground formations which store or carry pure water), major agricultural or grazing land, or forest land? If so, should the land be retained in its present use? Finally, is the land needed or well-suited for a different important use, such as housing, recreation, a school or hospital, or even another type of industry?*

A more detailed discussion of criteria for evaluating the choice of an industrial site and possible procedures for government regulation appears in a report by the American Bar Association called *Industrial Developments and the Environment: Legal Reforms to Improve the Decision-making Process in Industrial Site Selection —Review Draft* (Chicago, free). *Environmental Quality—1974*, mentioned above, includes a good discussion of basic land use principles, and gives a brief summary of each state's land use laws.

In addition to the above general considerations on site selection, you should also look at the specifics of site utilization, *e.g.*,

---

* Most of these general land use considerations were incorporated in the Jackson land use bill defeated in Congress in 1974.

Agricultural land near Lincoln, Nebraska.

location of production equipment, buffer zones, etc. Give particular attention to how the factory plans to dispose of solid and liquid wastes. An EIS or the company itself should be able to tell you about the nature and quantities of such expected wastes arising from production processes, air and water treatment systems, and from other plant sources, and about the proposed disposal site and precautions being taken to prevent adverse environmental effects. As mentioned in Chapter 2, you should be alert to any potential for escape of toxic substances into groundwater or in runoff.

In evaluating solid and liquid waste disposal plans, as well as the appropriateness and use of the site in general, you will probably want to consult specialists. A biologist or ecologist at a nearby university can help assess the effect of development upon wildlife. Geologists should be consulted if it appears that the proposed factory site may be subject to natural hazards such as floods or earthquakes. They can also evaluate long-term dangers of certain waste disposal practices. Local historians, anthropologists and archeologists can comment upon the factory's potential impact on historical or cultural values. Specialists from the local Soil Conservation Survey District (a division of the U. S. Dept. of Agriculture) can evaluate factory impact upon soil, water and other resources which contribute to the site's long-term productivity. Relevant specialists may be located at government agencies, such as the state land department, and state universities, which may have divisions specializing in resource evaluation.

AIR POLLUTION

A new plant's impact on air quality is one of the most important considerations to the surrounding community. You will thus want to obtain accurate estimates of emission levels, and evaluate the adequacy of planned control systems. You can find information on a proposed factory's potential air emissions in four places: in environmental impact statements (if prepared); in a company's own estimates of environmental impact; in an application for a construction permit filed with the state air pollution control agency; and in "new source reviews," which may be filed with the regional office of the U. S. Environmental Protection Agency.

As mentioned above, an EIS or a company assessment should contain a great deal of information on plant processes, pollution levels, and control systems. Similar information should be available prior to construction, from the state air pollution control agency, as Section 110 of the Federal Clean Air Act requires states to have a mechanism for preventing the construction or modification of an industrial plant which would result in violation of the Federal primary or secondary ambient air standards. As a result, most states require a company planning to build an industrial plant to file an application with the state air pollution control agency (see list, Chapter 3) for a construction permit. These applications contain extensive information on plant location, design, processes, pollution control equipment and expected emissions. (As the plant nears completion, the company must reapply for an operating permit, submitting additional information. This permit is usually granted, contingent on successful emissions testing after operations have begun.)

A fourth source of information may exist for new factories in industries for which the U.S. EPA has promulgated "New Source Performance Standards." Realizing that it is far easier to control pollution while building a new plant than to do so by adapting an old one, the drafters of the Clean Air Act specifically directed the EPA to develop special standards, reflecting the best system of emission reduction available for the industry, for new plants. As of early 1975, new source standards had been finalized for twelve industries, with standards for a dozen more scheduled to come out by the end of the year.

If the new plant belongs to one of these industries (see list, page 194), then the company may have petitioned the EPA for a "new source review," an informal evaluation of whether the new factory is likely to be able to meet the New Source Performance Standard. Since a new plant must perform emissions tests within two months after start-up, and must be in compliance with the standards within six months after beginning operation, a company may well choose to have its plans evaluated by the government beforehand.

Materials submitted as part of a new source review should be on file and publicly available at the regional EPA office (see list of addresses, Chapter 3), and should include fairly detailed information on process equipment specifications, air pollution control equipment, and projected emissions.

The kind of evaluation you can make of the proposed plant's air pollution will depend somewhat on its stage of development and the data you have been able to obtain. A good baseline figure for any evaluative effort is the plant's maximum "pollution potential." This can be determined as soon as you know the type and size of the production equipment the factory will use. As explained in Chapter 2 under "Pollution Potential," using figures in the EPA *Compilation of Air Pollutant Emissions Factors* and production figures, you can calculate how many pounds of each relevant pollutant the proposed plant would emit per day or year if it were to operate with no pollution controls whatsoever. (This is of course unlikely to be the case at a new plant.)

Hopefully you have been able to obtain other important data on the proposed plant as well, including projections of its actual emissions in pounds per hour of all the pollutants noted in the EPA *Compilation*, and the percent efficiencies expected from its pollution control equipment. If so, you can draw up a number of charts like those described in Chapter 4, which should help clarify how air pollution from the factory, as currently proposed, compares to legal standards, to "state-of-the-art" (the best achievable) control levels, to levels at other factories, and to levels adversely affecting public health and welfare.

As a first step, draw up a chart similar to Chart 2 in Chapter 4, comparing air emissions to state legal standards. Such standards are usually established in terms of pounds of pollutant allowable per hour from smokestacks. The factory's projected emissions of each pollutant in pounds per hour from each production process should be available from an environmental impact statement, an application for a construction permit from the state, or the company. The state standard, if one has been established, should be available from the state air pollution control agency.

If the new plant you are interested in belongs to one of the industries for which the EPA has promulgated New Source Performance Standards (NSPS's), you should draw up a second chart comparing air emissions to Federal legal standards. Most NSPS's are expressed in pounds of pollutant per pound or ton of product produced. If the company has filed for a "new source review," you should be able to obtain from this review projected air emissions expressed in these units. Otherwise, simply divide the emissions of each pollutant in pounds per hour by the tons of product to be made in the same time period.

The result is the factory's air pollution load in pounds per ton of product. (Even if you have obtained this statistic from other sources, you may still want to perform the calculation to check the figures for consistency. Inconsistencies do arise, due to data having been prepared at different times, or using different assumptions. Officials also sometimes make mistakes. Explanations of any inconsistencies should be sought from the sources of the data.) The final New Source Performance Standard should be obtainable from the regional EPA office, or from the *Federal Register* of the date on which it was promulgated (see list of dates).

Even if the proposed plant you are evaluating is not in one of the industries for which there are NSPSs, it is still useful to calculate its air pollution load in terms of pounds of each pollutant emitted per ton of product produced. This may then be used to construct a chart similar to Chart 3 in Chapter 4, comparing the new factory's emissions to the pollution load at other factories. You might choose, for such a comparison, another new factory which you know to be well controlled, using data obtained from the state air agency which regulates it.

If you have been able to obtain figures on the design efficiency of pollution control devices to be installed at the proposed plant (that is, the percent by weight of the pollutant in the stack gas which the devices will remove), then you can construct a chart similar to Chart 4 in Chapter 4, comparing the efficiency of the proposed plant's equipment to the efficiency of "state-of-the-art" systems—the best currently on the market. The projected efficiency figures for the factory may appear in an EIS or an application for a construction permit, or may be available from the company. The efficiency of the "best available" controls for these industrial processes should be gleaned from your background research into pollution control in the industry: reading of trade journals and EPA documents, interviews with the pollution control equipment manufacturers, etc.

Finally, you will want to clarify how the proposed plant will add to present air pollution levels in your area. This can be done in two ways.

AIR POLLUTION NEW SOURCE PERFORMANCE STANDARDS

Industry	Date Promulgated
Fossil Fuel-Fired Steam Generators	December 23, 1971
Incinerators	December 23, 1971
Portland Cement Plants	December 23, 1971
Nitric Acid Plants	December 23, 1971
Sulfuric Acid Plants	December 23, 1971
Asphalt Concrete Plants	February 28, 1974*
Petroleum Refineries	February 28, 1974*
Storage Vessels for Petroleum Liquids	February 28, 1974*
Secondary Lead Smelters	February 28, 1974*
Secondary Brass and Bronze Ingot Plants	February 28, 1974*
Iron and Steel Plants	February 28, 1974*
Sewage Treatment Plants	February 28, 1974*

Industry	Date Proposed
Primary Copper, Lead and Zinc Smelters	October 16, 1974
Ferroalloy Production Facilities	October 21, 1974
Steel Plants: Electric Arc Furnaces	October 21, 1974
Phosphate Fertilizer Plants	October 22, 1974
Primary Aluminum Plants	October 23, 1974
Coal Preparation Plants	October 24, 1974

* Amendments appear in the Federal Register dated March 8, 1974.

First, draw up a chart similar to Chart 6 in Chapter 4, noting projected factory emissions of each pollutant, and present total regional emissions from all sources. Your state air pollution control agency or the U.S. EPA National Air Data Bank at its Durham, N.C. center can provide the latter figures, as described in Chapter 3. Dividing factory by regional emissions and multiplying by 100 will give you the percentage by which the factory will increase the area's total pollution load.

While this will give you a sense of how much the new factory will contribute to your area's overall existing air pollution problems, it will not tell you how its impact will be distributed, or whether it will cause the air to become unhealthful. To determine this, the projected emissions from the smokestack must be used to calculate "ambient impact," a term denoting the contribution of factory emissions to general air pollution in the plant's vicinity.

The ambient impact of factory pollutants can be estimated only by meteorologists and other trained specialists, using complex mathematical models which simulate the manner in which soot, dust, and polluting gases disperse after leaving a factory smokestack.

However, these calculations may already have been done by the company in preparing an EIS, or the state air pollution control agency, from whom you could obtain the results. Try to obtain figures on both present ambient levels of various pollutants, and how much the new plant will increase them. Such figures are usually expressed in micrograms of pollutant per cubic meter of air (ug/m^3). Be particularly alert to whether pollution levels may be approaching the Federal primary and secondary ambient air standards (noted under "Effects of Pollution," Chapter 2).

Using the Data

The comparisons described above should show how factory air pollution may affect community health and welfare, whether the company is installing the best available pollution controls, and whether all existing legal standards will be met. Methods of evaluating your data and assessing pollution control strategies are fully described in the second half of Chapter 4. If you conduct an analysis as described there, you should be able to work with both regulatory and company officials to insure that the company complies with the full intent of the law, and preserves air quality to the maximum extent technologically and economically feasible.

In the area of government regulation, a new plant should of course meet all relevant standards, including the NSPS's and toxic substance limits. You might, however, want to turn particular attention to two areas. First, it is essential to ensure that there is no possibility that plant operation will cause Federal primary or secondary standards to be exceeded. The state air pollution control agency is not supposed to grant a construction permit unless it is certain that air quality will be preserved.* However, with so many factors affecting calculations of ambient impact— wind speed and direction, temperature, plume stability, etc.—the factory's potential effects on air pollution levels warrant your careful scrutiny.

If, for example, the environmental impact statement predicts that the new plant will raise particulate levels in your area by $15ug/m^3$ (micrograms per cubic meter) and the state air agency reports that current ambient particulate readings frequently rise to close to $45ug/m^3$, then the new plant may raise local particulate readings dangerously close to the Federal secondary standard of $60ug/m^3$. In such a case, you may want to call or arrange for additional review of ambient impact, either by the state, the company, or by independent authorities.

Second, you may want to assess whether the factory will cause "significant deterioration" of ambient air quality. This area of air pollution regulation, known as the "significant deterioration" provision, is one which is only beginning to be defined but which should have increasing importance and application in the future.

---

*The EPA has further stipulated that the standards be achievable using smokestacks not more than 2½ times the size of factory buildings. The factory may build taller smokestacks, but only to achieve even greater pollution reductions.

In 1971, when issuing ambient air standards, the EPA stated that, in accordance with the intent of the Clean Air Act, its issuance of these standards should not "be considered in any manner to allow significant deterioration of existing air quality in any portion of the state." Since then, the EPA has had to wrestle with exactly how to define this term.

Environmental groups have argued in court and elsewhere that the significant deterioration pledge requires EPA to ensure that states prohibit any new industrial development whose pollution would harm air quality in clean areas, even if that pollution would not cause the secondary standard to be violated. Industrial and developmental interests have countered that such a policy would effectively limit new factories to already polluted regions. Several EPA regulatory attempts in this area have been voided in certain court actions.

In late 1974, EPA issued a new set of regulations which it hopes will resolve some of the issues and stand up in court. The agency established a three-fold classification system under which a clean-air area may be designated either: a "Class I" area, where current low levels of sulfur dioxide and particulates must be maintained; a "Class II" area, where some changes in air quality through "well-controlled" development could take place; or a "Class III" area, where considerable industrial and other development could occur, provided sulfur dioxide and particulate levels do not exceed secondary standards. Under the system, all clean-air areas would start out with a Class II designation, and states would then have two years to propose redesignation to Classes I or III, subject to public hearings and EPA review. Unfortunately, as of early 1975, EPA was still only developing the criteria states would use in regulating development in Class II areas, and had not issued guidelines to be used in redesignations to I or III.

Prevention of "significant deterioration" is nevertheless still an important concept, and one whose application in your area should be considered, if you want to encourage compliance with the full intent of the law.

Even within the confines of the law, a company still frequently has a great deal of latitude in the manner and degree to which it controls air pollution. In discussions with company officials you may therefore want to give special attention to two points.

First, it is essential that toxic substance emissions be adequately controlled. As noted in Chapter 2, EPA is proceeding very slowly in its toxic substance control program and has issued standards for only asbestos, beryllium and mercury air emissions. You may therefore want to encourage the company itself to conduct a responsible and complete toxic substance testing and monitoring program, including thorough investigation of all possible carcinogenic or other disease-causing properties of new or untested chemicals it is planning to use in its processes, prior to any decision to use them. After plant start-up, it should institute continuous measurement of levels of dangerous substances in the workplace, and at points where such substances may escape from the plant. The offices of the National Institute for Occupational Safety and Health in Washington, D.C. and Cincinnati, Ohio, are presently the best central sources of information on factory toxic substance problems.

Second, you may want to encourage the company to utilize state-of-the-art controls to the extent feasible at the proposed factory. New Source Performance Standards officially mandate this level of performance. However, only a limited number of industries are now covered by NSPSs. Even in those industries, new, more effective pollution control methods are constantly being developed. Certain systems or process modifications may be possible at a particular plant which cannot be applied to the industry as a whole. Your charts on planned efficiencies of control devices and on emission levels at other new plants, as well as discussions with pollution control systems manufacturers, and review of trade journal materials, should help you evaluate what design improvements may be possible, and at what cost.

Indirect Sources: Vehicular Traffic

Besides process emissions, a new factory may contribute indirectly to air pollution by attracting vehicular traffic: cars driven to work by factory employees, trucks transporting raw materials or finished products, etc. Vehicular emissions, though generally

Air quality monitoring unit at an oil refinery.

not nearly as great as plant air pollution, may nevertheless warrant attention.

An analysis of the automotive pollution impact associated with the proposed factory you are researching may be available from the state air pollution control agency, which is responsible for maintaining ambient air standards, or from an EIS. The U.S. EPA has defined the circumstances under which, pursuant to the Clean Air Act, potential automotive emissions from indirect sources must be assessed They are: (1) if the proposed facility is to be located in a built-up area, as indicated by the area's inclusion among the U.S. Commerce Dept.'s Standard Metropolitan Statistical Areas (SMSAs) and the facility will have a parking lot holding 1000 or more cars; or (2) if it is to be outside an SMSA but will have a parking lot holding 2000 or more cars.

While the major industrial pollutants are particulates and sulfur dioxide, pollutants from autos are primarily hydrocarbons and carbon monoxide. (Nitrogen oxides may be generated by both kinds of sources.) In researching automotive pollution you should try to obtain two types of figures: potential absolute quantities of emissions, in pounds or tons per day or year; and the emissions' potential ambient impact, in parts per million to be added to the air at various locations near the factory. Both figures should be projected on a peak-hour, as well as an average basis, since rush hour traffic in or out of the plant is more likely to cause a violation of Federal one-hour ambient air standards.

You should compare projected ambient impact figures to current ambient air pollution levels and examine the potential for Federal primary and secondary standards to be exceeded. If you can obtain only absolute quantities of emissions, then you compare them to current emissions from all sources in the region (obtainable from the state air agency as noted above). If the projected emissions are large in comparison, you may want to request the state air agency or an independent analyst to determine their potential ambient impact.

If it appears that automotive emissions associated with the new factory may be significant you may want to encourage the company to take steps to reduce plant-related traffic. It can, for example, develop a low-fare bus system for commuting workers, devise an information system to facilitate employee car pooling, stagger work hours to reduce commuting during peak traffic hours, rely more heavily on rail or barge for freight movement, or even build a rail spur to the plant.

A regional or state transportation and planning agency may be able to assist you in pinpointing the alternatives most worth considering.

WATER POLLUTION

Since a new factory's water discharges can potentially affect recreational resources, drinking water purity, and vital ecological balances, it is important to analyze the nature of these discharges and their projected impact on water quality. Information on plant effluent and planned pollution control measures may be obtained from an environmental impact statement, from the company, and from applications for construction and discharge permits.

Although construction permit applications vary from state to state in their contents and timing, most should be available on request from your state's water pollution control agency (see list of addresses under "Legal Data," Chapter 3).

Comprehensive statistical data on the factory's projected pollution load should become available at least six months before the new factory is scheduled to begin operation, in the company's application for a National Pollution Discharge Elimination System (NPDES) permit. This will be filed with either the state water pollution control agency or the regional EPA office, or both. An NPDES application, described in Chapter 3, specifically requires information on factory water use, on each pollutant in the intake and discharge water, and on control equipment.

Using the information collected from these sources, you can begin evaluating the new factory's projected water pollution load. First, organize the data along the lines described in the "Water Pollution" section of Chapter 4. Using figures from the NPDES application or similar statistics gathered from the other sources, you should be able to draw up charts similar to Chart 7, comparing the factory's proposed discharge rate in gallons per day to the flow of the receiving waterway; Chart 8, comparing projected concentrations in parts per million of each pollutant to the concentrations at which adverse environmental effects are known to occur; and Chart 12, comparing the design efficiency of the new factory's planned water pollution control equipment to the efficiency of state-of-the-art control systems. (See Chapter 4 for a step-by-step description of how to construct these charts.)

If the proposed plant is in one of the 40 industries for which the EPA has developed effluent guidelines (see list, page 14), you should draw up a chart similar to Chart 10 in Chapter 4. However, instead of comparing the factory to the BAT and BPT (Level I and II) control levels developed for existing plants, its projected discharges in pounds per ton of product should be compared to the "New Source" of NSPS (Level III) guidelines for the industry. These guidelines are designed to reflect the pollution control capabilities of a plant operating not only with the best available pollution control equipment, but also with the cleanest production processes and operating procedures. NSPSs are either as strict as or stricter than the BAT (Level II) guidelines which existing plants must achieve by 1983.

The new factory's projected discharges of various pollutants in pounds per ton of product should be calculated as explained for Chart 10. Proposed NSPS guidelines can be obtained from the *Development Document* for the industry. The final, promulgated guidelines can be obtained from the *Federal Register* of the dates on which they were issued, or from the regional EPA office.

If there are no NSPSs for the industry, you can, if you have the data, compare the factory's proposed discharges per ton of product to the discharges of a very well-controlled factory which makes the same product.

Finally, you should attempt to draw up a chart similar to Chart 11 in Chapter 4 comparing the new factory's projected water discharges to the actual legal limits which it must meet. If legal limits have been established, they will appear in the NPDES plant permit, which is officially granted by either the regional EPA office or the state water pollution control agency. Several sets of legal standards must be complied with.

A new plant must meet the industry New Source Performance Standards, if they were promulgated prior to the start of factory construction. It must also comply with any toxic substance standards. (As explained in Chapter 2, EPA has made a list of 123 toxic water pollutants, proposed standards for nine, but as of early 1975 had promulgated none; however, such standards may exist in the future.)

In addition, the factory cannot cause a violation of "water quality standards." Laws passed by Congress prior to the 1972 Water Pollution Control Act required each state to categorize its interstate lakes, rivers, streams and coastal waters according to desired usage, *i.e.*, for swimming, drinking, fishing or industrial use, and to determine the maximum amounts of pollution which each water body could contain without becoming unsuitable for that use. This system of standards was incorporated in the 1972 Act and extended to apply to all intrastate waters.

Each state is now required to predict the impact of a new factory's water discharges on the receiving water body. The evaluation may be done by the state water pollution control agency, or by special area-wide waste management planning agencies called for under the 1972 Act. If the agency concludes that water quality standards will be violated (even though the factory may meet EPA effluent guidelines), then it must establish legal limits for the plant to prevent violation.

A final possible step in assessing new factory water pollution is to estimate its effect on pre-existing pollution levels. This can be done through mathematical models which predict the amount by which factory effluent will be diluted at various points downstream. While such calculations should not be undertaken without the help of a trained water pollution specialist, such an analysis may have already been performed and be available in an environmental impact statement, from the company, or from the state water pollution or area-wide planning agency.

Using the Data

The data you have collected and organized on factory water pollution should allow you to draw significant conclusions about how plant discharges compare to legal standards, to "state-of-the-art" control levels, to levels having adverse environmental effects, and to performance elsewhere in the industry. Methods of analyzing the data are fully discussed in the second half of Chapter 4.

In discussions and interactions with regulatory and company officials about potential water pollution problems, there are certain areas to which you should devote particular attention.

The first is the potential environmental effects of factory water pollutants. The state is supposed to insure that the receiving waterway's "designated use" will not be affected by factory discharges. However, the assumptions and models behind its assessment may be open to interpretation, and you may want to obtain from qualified experts an outside opinion on the agency's evaluation.

You should also give particular attention to discharges of pollutants which are known to be, or suspected of being, toxic, such as heavy metals and organic chemicals, since this is an area in which standards are not yet well-developed. You should refer to your chart on projected pollution concentrations in factory effluent, to any data obtained on ambient levels downstream, and to the research sources on effects of water pollutants described in Chapter 2, for evidence of potentially harmful discharges. This is especially important in terms of effects on human health if the water body is a source of drinking water, or is heavily used for fishing or recreation.

Second, although use of state-of-the-art control is legally mandated only for factories on which work has begun after the issuance of final New Source Performance Standards for the industry, you may nevertheless want to encourage use of such controls at a new factory not covered by NSPSs. This could be to the company's, as well as the community's, advantage, since it would give it a "head start" in meeting the standard to be applied in 1983—the date by which all plants, regardless of industry or date of construction, must be using the "best available" controls. Your charts on efficiency of water pollution controls and discharge rates in pounds per ton of product at other factories should help you evaluate whether the new factory's systems will be state-of-the-art.

Water quality monitoring at oil refinery.
Aeration cone provides secondary treatment.

A third area which may be of concern in certain parts of the country is a proposed factory's consumption of water: not the amount taken into the plant and discharged again, but the amount "used up," through evaporation or other factory processes. A large, 1000-megawatt nuclear power plant, for example, will evaporate roughly 12 million gallons of water during a typical day's operations. In preparing your chart on water use, you should have been able to determine whether the proposed factory's water consumption (intake minus discharge) will be significant.

The document in which all factory legal requirements will be incorporated is the NPDES permit. If the factory has not yet received one, you should be sure to take advantage of opportunities for public participation in the permit-granting process. This and other approaches to conveying your environmental concerns and conclusions to company officials and regulators are discussed in Chapter 5.

Indirect Sources: Construction

Besides discharging pollution in its effluent, a new factory can contribute indirectly to water pollution, particularly while being built. The construction of an industrial plant can cause special environmental problems, of which erosion, with its impact on water and land resources, is typically the most serious. With natural vegetation disturbed and topsoil exposed, dirt, oil, chemicals and other substances may wash from the site and muddy and pollute lakes and streams.

The 1972 Federal Water Pollution Control Act deals with pollution from "non-point" sources, including construction sites, in Section 208. This part of the law requires each state, or another designated planning agency, to set up area-wide waste management planning programs, whose functions include identifying construction-related sources of water pollution and establishing procedures and methods for controlling them "to the extent feasible." However, as of March, 1975, the 52 planning agencies which had been "designated" were only beginning to receive funding to develop their programs, indicating that this part of the law is only in the very earliest stages of enforcement. You can determine if there is a designated planning agency in your area by consulting the state water agency or the U. S. EPA. The planning agency, if it exists, should be able to tell you whether it has developed any guidelines or regulatory mechanism to deal with construction.

You may also be able to obtain information on construction plans from the company, or from an environmental impact statement. Measures to prevent erosion from construction sites, fully described in the literature of soil conservation and civil engineering, are well-summarized in an EPA report entitled *Processes, Procedures and Methods to Control Pollution Resulting from All Construction Activity* (GPO #EPA-430/9-73-007, $2.30). They include grading and revegetation to minimize the extent of soil surface exposed at any given time; using drainage ditches to divert and control runoff; and using mulch to shield soil from rain and runoff. The most appropriate erosion control program for a site will depend upon such factors as contour, soils and rainfall, as well as construction requirements.

Since some erosion will result from construction, even with a well-conceived prevention program, construction plans should also provide for preventing eroded soil from ending up in waterways. Sediment control practices include vegetative buffers between an exposed graded slope and a waterway; sod or gravel filters to detain runoff; settling basins; and ponds to hold stormwater.

Besides sediment dislodged through erosion, other substances dispersed from construction sites may ultimately contribute to pollution of lakes and streams. These may include herbicides and pesticides; fuels, lubricants and asphalt; paints, acids and cleaning solvents; waste water from cooling and mixing concrete; and leftover cement, lime, fertilizer and garbage. Many of these substances unfortunately cannot be adequately controlled by erosion and sediment control methods, especially those which are soluble in water, or which form surface films on water (*e.g.*, oil and grease).

At present, the most widely-used approach for controlling these pollutants is to exercise "good housekeeping": using only required amounts of potentially-polluting substances, applying them properly, and avoiding careless mistakes or spills in which large quantities of pollutants are dumped onto the land.

Monitoring by environmental officials or citizen groups—if that can be arranged—may be the only available way to encourage practices which could lead to minimizing pollution from construction. Experts in the field generally agree that to curb construction-related pollution effectively, control strategies must be spelled out in the contracts let by the company for design and construction of the new facility. You might try to arrange with the company for permission to examine construction contracts to ensure that provision has been made for controlling erosion, sedimentation and pollutant runoff. The Soil Conservation Service of the U.S. Agriculture Department, county soil and water conservation districts, and the U.S. Geological Survey may have reports on soil composition, topography, water resources, and other important factors relating to the facility's site. Officials of these agencies are in a position to interpret information in these reports for you, and in some instances may be willing to work with you in monitoring the company's program to minimize construction pollution.

ENERGY USE

All factories use energy, in the form of electricity, or through burning fuels, or both. Since present methods of producing energy invariably result in adverse environmental effects, ranging from combustion-caused air pollution to losses in agricultural productivity of strip-mined lands, and since energy is an increasingly scarce resource in this country, you should examine carefully the proposed factory's energy use and how it can be minimized.

A new factory's projected fuel use (and resulting air emissions) —whether the fuel is to be burned to produce steam or hot air for industrial processes, to run motors and assembly lines, to generate electricity on-site, or simply to heat the workplace—is generally included in any environmental impact statement, and in the application to the state air pollution control agency for a construction permit.

Projected electricity use (and the pollution resulting from power generation) is generally *not* included in environmental assessments, when the power is to be purchased from an electric utility.

You should, therefore, attempt to determine how much electricity the proposed plant will use, where it will be obtained, and how much pollution will be caused by its generation, since such power purchases may seriously affect both regional power supplies and air pollution levels.

The best source of data on the amount of electricity the plant will use is, of course, the company building the factory or the supplying utility (if you know it). You should request the following information:

(1) Projected annual electricity usage in megawatt hours (mw-hrs). A megawatt-hour is one thousand kilowatt hours.

(2) Projected "peak" electrical demand in megawatts, reflecting the rate at which the factory will use power when running at top capacity.

(3) Projected daily and annual operating cycle; that is, the times during the day or year when periods of peak operation are expected to occur.

If company or utility data is unavailable, you can make an extremely rough estimate of the plant's future electric demand, based on the demand of existing factories in the industry, using sources of information described in Chapter 2 under "Factory Raw Material Use" and "Factory Energy Use."

Once you have an estimate of the proposed plant's annual electricity consumption, you should try and find out the annual sales, in megawatt-hours, of the utility which will be supplying the plant. This figure can be obtained from the utility.

If you divide the proposed plant's annual usage by the utility's annual sales, both expressed in megawatt-hours, and multiply by 100, the result is the percent by which the new plant would increase present sales, a relatively simple, rough way of visualizing the relative magnitude of its electrical demand on the utility.*

---

* If the electric utility serving the new factory operates a predominantly fossil-fired system, then the approximate *fuel* consumption incurred in supplying the factory's electricity can be computed by multiplying the number of mw-hrs. purchased, times: 0.4, to obtain the number of tons of coal burned, or 1.6 to obtain the number of barrels of oil burned; or 10, to obtain the number of "mcf" (thousands of cubic feet) of natural gas burned.

If the factory will make relatively low demands on the utility—in general, increasing sales by less than 1%—then you probably need not concern yourself with its electricity use any further, although you should still encourage basic energy conservation, as noted below. However, if the factory's projected use is significant relatively to the utility's current output, you may want to make an assessment of the new factory's potential effect on local power supplies and on the utility's environmental impact.

Crucial to such an assessment is the timing of the electric utility's and the proposed factory's peak demand periods. The utility should be willing to supply you with this data on its operations, which can then be compared to company peak-use data. If the factory's peak demand will coincide with the overall peak demand exerted by the utility's other customers, this could lead to power shortages; while if the factory's peaks occur during times of low demand, existing equipment might be able to supply the factory. One way to avoid peak demand problems might be to charge the factory higher rates for electricity consumed during peak-use periods.

You may also, in considering the new factory's total environmental impact, want to consider the amount of air pollution which will result from meeting its electricity demands. One approach to estimating this is to extrapolate on the basis of the utility system's present emissions.

Begin by adding up the pounds of particulates emitted per hour at each of the utility's generating plants. The best source of this information is state air pollution agencies, and the FPC, as noted in Chapter 3. An annual Federal Power Commission publication called *Steam-Electric Air and Water Quality Control Data* (available from the FPC office in Washington, D. C., $2.00) also includes power plant emission data, expressed in tons per year, although its usefulness is limited by the fact that the statistics in it are approximately three years old when published. Repeat the addition process for sulfur dioxide and nitrogen oxide emissions.

If you then multiply the total for each pollutant by the percent by which the new factory will increase demand, calculated above, you will have a very rough estimate, in either pounds per hour or tons per year, of the power plant air pollution the new factory may cause. For example, if an electric utility has four coal-burning power plants which emit 2000, 3000, 7500, and 4300 pounds of particulate matter an hour, respectively, the utility's total particulate emissions are 16,800 pounds an hour. If the proposed plant's projected power use amounts to 5 per cent of current demand, then it may raise particulate air emissions by 16,800 × .05 = 840 pounds per hour.

The pollution problems associated with both nuclear and fossil-fuel-fired power plants, and methods of alleviating them, are described in detail in a report, titled *The Price of Power*, produced by the Council on Economic Priorities, a New York-based public interest group (MIT Press, Cambridge, Mass. $18.50).

A final very important way to avoid the environmental impact of power generation and to save fuels as well, is to conserve energy. In the case of a new factory, it is particularly important to consider energy conservation measures in the designing and planning stages, since many measures—such as the structuring and insulating of a building to retain heat in the winter and to stay cool in summer, or the selection of most efficient motors—are either difficult or impossible to incorporate later. Energy conservation methods for the iron and steel, petroleum, paper, aluminium, copper and cement industries are discussed in some detail in a Ford Foundation-sponsored study called *Potential Fuel Effectiveness in Industry* (Ballinger Publishing Co., Cambridge, Mass. $2.50). A checklist of energy conservation measures for small to intermediate-sized manufacturing firms appears in a National Bureau of Standards handbook entitled, *Energy Conservation Program Guide for Industry and Commerce* (GPO C13.11:115, $2.50).

Ultimately, you may want to encourage the company to employ an engineer or consultant with expertise in energy conservation to review projections of fuel and energy consumption, and determine whether all available, economically justifiable energy conservation measures are being incorporated into the proposed plant's design. In the present context of rising energy prices, this may ultimately be in the best interests of both the company and the community. A seven-page list of offices and organizations through which energy expertise can be sought appears in the *Energy Conservation Program Guide* mentioned above.

SOCIAL IMPACT

The construction and operation of new plants can exert a major and lasting impact on the life and environment of nearby communities, quite apart from strictly ecological effects on air, water and land. These "social impacts," most notably on population growth, may be of great concern to present residents. In understanding and evaluating a new factory, you may want to consider its potential effects on the "social" environment.

The establishment of new factories often occurs as a natural and necessary part of the economic evolution of a community or region. As population increases, and as some jobs are phased out by the forces of competition and technological change, new plants are needed to provide jobs and sustain a community's economic vitality. Accordingly, state and local governments—as well as business and labor groups—have traditionally worked to attract new industry to their area. Environmentalists, while often critical of these groups for alleged lack of concern for environmental values, generally recognize the need to tolerate *some* pollution from new plants, as the price of maintaining full employment in the community.

Sometimes, however, industrial development does not fit this evolutionary pattern. A case in point is the proposal of energy companies and Federal agencies to open coal-rich areas of four Northern Plains states to large-scale strip mining. The bulk of the coal produced would feed massive power plants and coal gasification facilities to be built at the mines, with electricity and gas to be shipped to West Coast and Midwest population centers. Considerable mining and burning of coal is already under way, and some projections show the region producing more electricity by the year 2000 than any *country* in the world does now except the United States and the Soviet Union.

Such development would create a major and permanent upheaval in the lives of people in the region, according to the September,

Transmission lines near Searchlight, Nevada.

Strip mining on Indian burial grounds, Black Mesa, Arizona.

1974 *Northern Great Plains Resource Program Draft Report,* a joint state and Federal study. In the "principal impact areas," within the four states, coal development under a "rapid development profile" posited in the study would cause the population to more than double—from 434,000 in 1970 to 950,000 in the year 2000—compared to population growth of only one percent over the 1960's. Employment would increase by 143 percent.

The study expresses concern over the capability of local institutions to provide housing, education, health care and municipal services to such rapidly expanding populations. While development generally adds significantly to a region's tax base, tax revenues often do not begin growing until several years after the construction crews and factory workers and their families have arrived with their need for government services.

Changes similar to those being predicted as a consequence of rapid coal development can be expected in many other areas which will be newly industrialized by a major factory. If you are concerned about the possible social impact of a proposed industrial development, there is some basic research you should do.

First, consider your local economic situation. If the community or region is characterized by a stable economy with low unemployment, then new development may require substantial in-migration of a labor force. Perhaps a different location for the plant should be considered.

While some industries, especially those which extract or process raw materials, have relatively little flexibility in where they locate, more socially desirable alternatives may still be available. For example, opponents of Western energy development argue that mining companies could develop additional coal resources in Appalachia, which has high unemployment, rather than bring disruptive development to the West.

Second, consider the size and type of the labor force, including the construction workers, required. Obviously, when fewer workers are involved, less social impact will be generated by in-migration or temporary residence.

Third, consider the time frame involved in building the factory. The greater the time period between the initial proposal for the new plant and the beginning of construction and operation, the greater the community's ability to plan for social change, such as new patterns of land use and housing, and new facilities for education, health care, recreation, water, sewage and sanitation. Consider also how long the industry will remain. A mining operation may exhaust an ore body within 10 to 20 years, leaving significant social and environmental disruption and little economic benefit behind.

Finally, examine possible "spin-offs" from development, such as how the new plant may affect existing ways of making a living or whether it may attract other industries. For example, those opposed to Northern Plains coal development have argued that water pollution and consumption for coal mining, burning and gasification will reduce the availability of water for agriculture, threatening the livelihoods of ranchers and farmers. On the other hand, fertilizer plants might be built near coal gasification units, to utilize by-products of the process, which farmers might consider beneficial.

If your research suggests that building and operating the new facility will bring significant disruption to the local community or region, then you have strong grounds for petitioning government to conduct a comprehensive study of its social impact. Your state government is probably best equipped for this task, although a strong county planning agency or regional planning association may also be well suited. If the proposed development is truly enormous in scope, or involves a particular area of national significance (such as an historic or scenic region), then participation by a Federal agency, such as the Department of Interior, may be appropriate.

Any social impact study should assess: the size and type of labor force required; the expected amount of population in-migration, the probable places of settlement and housing requirements of new residents, the increase in demand for community and institutional services such as sanitation, police and fire protection, health care, and education; impact upon current livelihoods; and the prospects for desirable or adverse "spinoff" development.

References

The research materials referred to in this guide, grouped here for convenience with full bibliographical information and addresses where they can be ordered, fall into several categories. First there are books and pamphlets. Some you may want to purchase, others refer to in a library. Second are periodicals: you may want to subscribe to some, read back issues of others in a library, and with still others, read only articles noted in indexes. Indexes, the third category, are generally either available free on request or are available in a library. The fourth category is data sources, reports which contain important information but which are neither books and pamphlets, periodicals nor indexes.

The references are arranged according to chapter where discussed. The title, publisher, city of publication, price, and in the case of government publications, the GPO or NTIS number are noted in the text of the chapter, as well as here.

Orders to the GPO and NTIS must be accompanied by payment in advance. Their addresses are as follows:

Superintendent of Documents
Government Printing Office
Washington, D.C. 20402

National Technical Information Service
5282 Port Royal Rd.
Springfield, Va. 22151

CHAPTER 2

Books

A PRIMER ON WASTE WATER TREATMENT, U. S. Environmental Protection Agency. Available from Superintendent of Documents, Government Printing Office, Washington, D. C. 20402, GPO #0-419-407, $.55.

AIR POLLUTION ASPECTS OF (NAME OF POLLUTANT), U. S. Dept. of Health, Education and Welfare, 1969. Available from National Technical Information Service, 5282 Port Royal Rd., Springfield, Va. 22151, NTIS # varies (see list, Chapter 2), $6.00.

AIR POLLUTION ENGINEERING MANUAL, Second Edition, compiled and edited by John Danielson, U. S. Environmental Protection Agency, 1973. Available from Superintendent of Documents, Government Printing Office, Washington, D. C. 20402, GPO #4.9:40/2, $14.50.

AIR POLLUTION PRIMER, National Tuberculosis and Respiratory Disease Association, 1969. Available from the American Lung Association, 1740 Broadway, New York, N. Y., free.

AIR QUALITY CRITERIA FOR (NAME OF POLLUTANT), U. S. Dept. of Health, Education and Welfare, 1969. Available from Superintendent of Documents, Government Printing Office, Washington, D. C. 20402, or National Technical Information Service, 5282 Port Royal Rd., Springfield, Va. 22151, order # varies (see list Chapter 2), $1.50-$1.75.

BACKGROUND INFORMATION FOR PROPOSED NEW SOURCE PERFORM-ANCE STANDARDS, U. S. Environmental Protection Agency. Available from the National Technical Information Service or the U. S. Environmental Protection Agency, order # varies (see list Chapter 2), 0-$4.50.

CENSUS OF MANUFACTURERS (1972), U.S. Dept. of Commerce, 1975. Available in business libraries; industry segments available from the Superintendent of Documents, Government Printing Office, Washington, D. C. 20402, GPO # varies (see Chapter 2), $.75–$1.40.

CLEANING OUR ENVIRONMENT: THE CHEMICAL BASIS FOR ACTION, edited by Lloyd M. Cook, ACS Committee on Chemistry & Public Affairs, 1969. Available from the American Chemical Society, 1155 16th Street, NW, Washington, D. C. 20036, $2.75.

COMPILATION OF AIR POLLUTANT EMISSION FACTORS, 2nd EDITION, U. S. Environmental Protection Agency, 1973. Available from Superintendent of Documents, Government Printing Office, Washington, D. C. 20402, GPO #EP4.9:42/2, $2.90.

CONTROL TECHNIQUES FOR (NAME OF POLLUTANT), U. S. Dept. of Health, Education & Welfare, 1969. Available from Superintendent of Documents, Government Printing Office, Washington, D. C. 20402, GPO # varies (see list Chapter 2), $.70-$2.10.

CONTROLLING AIR POLLUTION, Rena Corman, American Lung Association, 1974. Available from the American Lung Association, 1740 Broadway, New York, New York, free.

COST OF AIR POLLUTION DAMAGE: A STATUS REPORT, Larry Barrett & Thomas Waddell, U. S. Environmental Protection Agency, 1973. Available from Superintendent of Documents, Government Printing Office, Washington, D. C. 20402, GPO #EP4.9:85, $.70.

CRITERIA FOR A RECOMMENDED STANDARD FOR OCCUPATIONAL EXPOSURE TO (NAME OF POLLUTANT), National Institute for Occupational Safety and Health. Available from Superintendent of Documents, Government Printing Office, Washington, D. C. 20402, GPO # varies (see list Chapter 2), $.90-$2.10.

DEVELOPMENT DOCUMENTS FOR EFFLUENT LIMITATION GUIDELINES AND NEW SOURCE PERFORMANCE STANDARDS, U. S. Environmental Protection Agency. Available from Superintendent of Documents, Government Printing Office, Washington, D. C. 20402, GPO # varies (see list Chapter 2), $1.60-$20.30.

ECONOMIC ANALYSIS OF PROPOSED EFFLUENT GUIDELINES, U.S. Environmental Protection Agency. Available from National Technical Information Service, 5282 Port Royal Rd., Springfield, Va. 22151, NTIS # and price varies (see list, Chapter 2).

ENERGY CONSUMPTION IN MANUFACTURING, John G. Myers, Project Director, The Conference Board, 1974. Available from Ballinger Publishing, 17 Dunster Street, Harvard Square, Cambridge, Massachusetts 02138, $9.95.

ENVIRONMENT USA, compiled and edited by the Onyx Group, 1974. Available from Xerox Corporation, R. R. Bowker, PO Box 1807, Ann Arbor, Michigan 48106, $15.95.

ENVIRONMENTAL QUALITY—1973: THE FOURTH ANNUAL REPORT OF THE COUNCIL ON ENVIRONMENTAL QUALITY, The President's Council on Environmental Quality. Available from Superintendent of Documents, Government Printing Office, Washington, D. C. 20402, GPO #4111-00020, $4.30.

ENVIRONMENTAL STEEL, James Cannon, CEP, 1973, Available from Praeger, 111 Fourth Ave., N.Y.C., $18.50.

FUNDAMENTALS OF AIR POLLUTION, 3 vols., Arthur Stern, Academic Press, 1973. Available from Academic Press, 111 Fifth Avenue, New York, N. Y., $14.50.

INDUSTRIAL POLLUTION CONTROL HANDBOOK, H. F. Lund, McGraw-Hill, 1971. Available from McGraw-Hill, 1221 Avenue of the Americas, New York, N. Y. 10022, $29.50.

INDUSTRY EXPENDITURES FOR WATER POLLUTION ABATEMENT, Leonard Lund, The Conference Board, 1972. Available from the Conference Board, 845 Third Avenue, New York, N. Y. 10022, Conference Board Report #541, $3.50 for Associates and educational groups, $17.50 for others.

NATIONAL ENVIRONMENTAL RESEARCH CENTER PROGRAM DIRECTORY, prepared by Office of Program Management, of the U. S. Environmental Protection Agency, 1974. Available from Office of Program Management, Office of Research and Development, U. S. Environmental Protection Agency, Washington, D. C. 20460, free.

PERMITS FOR WORK AND STRUCTURES IN, AND FOR DISCHARGES OR DEPOSITS INTO NAVIGABLE WATERS, U. S. Army Corps of Engineers, 1971. Available from U. S. Army Corps of Engineers, Washington, D. C. or U. S. Environmental Protection Agency regional offices, free.

POTENTIAL FUEL EFFECTIVENESS IN INDUSTRY, Elias Gyftopoulos, Thermo Electron Corporation, 1974. Available from Ballinger Publishing Company, c/o J. B. Lippincott, East Washington Square, Philadelphia, Pennsylvania, paper—$3.00, hard bound—$8.00.

STANDARD INDUSTRIAL CLASSIFICATION MANUAL, the President's Office of Management and Budget, 1972. Available from Superintendent of Documents, Government Printing Office, Washington, D. C. 20402, GPO #4101-0066, $6.75.

THE ECONOMIC IMPACT OF POLLUTION CONTROL: A SUMMARY OF RECENT STUDIES, U. S. Environmental Protection Agency, 1972. Available from Superintendent of Documents, Government Printing Office, Washington, D. C. 20402, GPO #0-458-471, $2.50.

THRESHOLD LIMIT VALUES FOR CHEMICAL SUBSTANCES AND PHYSICAL AGENTS IN THE WORKROOM ENVIRONMENT WITH INTENDED CHANGES FOR 1973, American Conference of Governmental Industrial Hygienists, 1973. Available from Secretary-Treasurer, American Council of Governmental Industrial Hygienists, PO Box 1937, Cincinnati, Ohio 45201, $.75.

WATER QUALITY CRITERIA, Report of the National Technical Advisory Committee to the Secretary of the Interior, April 1, 1968, Federal Water

Pollution Control Administration (reprinted by the U. S. Environmental Protection Agency, 1972). Available from Superintendent of Documents, Government Printing Office, Washington, D. C. 20402, GPO #EP1.23:73-033, $12.80.

WATER QUALITY CRITERIA DATABOOK, VOLUME I: ORGANIC CHEMICAL POLLUTION AND VOLUME II: INORGANIC CHEMICAL POLLUTION, U. S. Environmental Protection Agency, 1970. May be available in environmental libraries; reference no. EPA 180-10DPV12/70.

WATER QUALITY TRAINING INSTITUTE INFORMATION KIT, available from The Conservation Foundation, 1717 Massachusetts Avenue, NW, Washington, D. C. 20036, $10.00, $6.00 to non-profit groups.

Periodicals

AIR/WATER POLLUTION REPORT, weekly, Business Publishers Inc., Box 1067, Blair Station, Silver Spring, Maryland 20910, $145/year.

BUSINESS WEEK, weekly, McGraw-Hill, McGraw-Hill Building, 1221 Avenue of the Americas, New York, N. Y. 10020, $17.00/year.

CHEMICAL ENGINEERING, bi-weekly, McGraw-Hill, McGraw-Hill Building, 1221 Avenue of the Americas, New York, N. Y. 10020, $12.00/year.

ENVIRONMENT, monthly, Subscriptions, Box 755, Bridgeton, Missouri 63044, $10.00/year.

ENVIRONMENTAL SCIENCE & TECHNOLOGY, monthly, American Chemical Society, Subscription Service Department, 1155 16th Street, NW, Washington, D. C. 20036, $9.00/year.

EPA CITIZENS' BULLETIN, monthly, U.S. EPA Office of Public Affairs, Washington, D.C., free.

FORBES, monthly, Forbes, Inc., 60 Fifth Avenue, New York, New York 10011, $12.00/year.

FORTUNE, monthly, Time, Inc., 541 North Fairbanks Court, Chicago, Illinois 60611, $14.00/year.

JOURNAL OF AIR POLLUTION CONTROL ASSOCIATION, Air Pollution Control Association, 4400 5A, Pittsburgh, Pennsylvania 15213, $25.00/year.

JOURNAL OF WATER POLLUTION CONTROL FEDERATION, Water Pollution Control Federation, 3900 Wisconsin Avenue, Washington, D. C. 20022, $22.00/year.

MONTHLY ECONOMIC LETTER, First National City Bank, 399 Park Avenue, New York, N. Y. 10022, free.

NEW YORK TIMES, daily, 229 West 43rd Street, New York, New York 10036, $114.00/year mail delivery, $144.00/year home delivery.

NEWSWEEK, weekly, Subscriber Service, 444 Madison Avenue, New York, N. Y. 10022, $19.50/year.

SCIENCE, weekly, American Association for the Advancement of Science, 1515 Massachusetts Avenue, NW, Washington, D. C. 20005, $50.00/year.

SCIENTIFIC AMERICAN, monthly, 415 Madison Avenue, New York, N. Y. 10017, $15.00/year.

TIME MAGAZINE, weekly, Time, Inc., 541 North Fairbanks Court, Chicago, Illinois 60611, $18.00/year.

WALL STREET JOURNAL, daily, except Sat. and Sun., 22 Cortlandt Street, New York, New York, $42.00/year.

Indexes

AIR POLLUTION TECHNICAL PUBLICATIONS OF THE U. S. ENVIRONMENTAL PROTECTION AGENCY, published semi-annually by the Air Pollution Technical Information Center (APTIC) of the Environmental Protection Agency, Washington, D. C. Available from APTIC, Research Triangle Park, North Carolina, free.

APPLIED SCIENCE & TECHNOLOGY INDEX, H. W. Wilson & Co., available at large libraries.

AVAILABLE INFORMATION MATERIALS, Solid Waste Information Materials Control Agency, Cincinnati, Ohio, free.

BIBLIOGRAPHY OF R & D RESEARCH REPORTS, Office of Research & Development, U. S. Environmental Protection Agency, Washington, D. C. 20460, subscriptions free.

BUSINESS PERIODICAL INDEX, H. W. Wilson & Co., available at large libraries.

CURRENT PUBLICATIONS, National Institute for Occupational Safety and Health, Office of Technical Publications, Room 530, Cincinnati, Ohio, free.

ENCYCLOPEDIA OF ASSOCIATIONS, VOL. I: NATIONAL ORGANIZATIONS IN THE UNITED STATES, Gale Research Company. Available from the Gale Research Company, Detroit, Michigan, $55.00, and business libraries.

FUNK & SCOTT INDEX OF COMPANIES AND INDUSTRIES, Funk & Scott, Inc., available at large libraries.

READER'S GUIDE TO PERIODICAL LITERATURE, H. W. Wilson & Co., available at large libraries.

RESEARCH REPORTS, published quarterly by the Office of Water Resources Research of the U. S. Department of the Interior, available from Water Resources Scientific Information Center, U. S. Department of the Interior, Washington, D. C., free.

SELECTED WATER RESOURCES ABSTRACT, available from Water Resources Scientific Information Center, U. S. Department of the Interior, Washington, D. C., free.

Data Sources

ANNUAL REPORTS, available from the company.

NATIONAL POLLUTION DISCHARGE ELIMINATION SYSTEM PERMIT APPLICATIONS, available from U. S. Environmental Protection Agency regional offices.

REFUSE ACT PERMIT APPLICATIONS, available from U. S. Environmental Protection Agency regional offices.

10K, 8K FORMS, available at regional SEC offices.

CHAPTER 3

Books

A CITIZEN'S GUIDE TO CLEAN AIR, The Conservation Foundation, 1972. Available from The Conservation Foundation, 1717 Massachusetts Avenue, NW, Washington, D. C. 20036, free.

A CITIZEN'S GUIDE TO CLEAN WATER, Izaak Walton League of America, 1973. Available from the Izaak Walton League of America, Suite 806, 1800 North Kent Street, Arlington, Virginia 22209, free.

ENVIRONMENTAL STEEL: POLLUTION IN THE IRON AND STEEL INDUSTRY, James Cannon, Council on Economic Priorities, 1973. Available from Praeger Publishers, Inc., 111 Fourth Ave., New York, N. Y. 10036, $18.50.

HOW TO CHALLENGE YOUR LOCAL ELECTRIC UTILITY: A CITIZEN'S GUIDE TO THE POWER INDUSTRY, Sandra Jerabek & Richard Morgan, Environmental Action Foundation, 1974. Available from the Environmental Action Foundation, 720 Dupont Circle, Washington, D. C. 20036, $1.50.

PAPER PROFITS: POLLUTION IN THE PULP AND PAPER INDUSTRY, Leslie Allan, Eileen Kohl Kaufman, Joanna Underwood, Council on Economic Priorities, 1970. Available from MIT Press, 28 Carlton Street, Cambridge, Massachusetts 02142, $20.00.

PERMITS FOR WORK AND STRUCTURES IN, AND FOR DISCHARGES OR DEPOSITS INTO NAVIGABLE WATERS, U. S. Army Corps of Engineers, 1971. Available from U. S. Army Corps of Engineers, Washington, D. C. or U. S. Environmental Protection Agency regional offices, free.

PRICE OF POWER: ELECTRIC UTILITIES AND THE ENVIRONMENT, Charles Komanoff, Holly Miller, Sandy Noyes, Council on Economic Priorities, 1972. Available from MIT Press, 28 Carlton St., Cambridge, Mass. 02142, $18.50.

TOWARD CLEANER WATER: THE NEW PERMIT PROGRAM TO CONTROL WATER POLLUTION, Environmental Protection Agency, 1974. Available from Superintendent of Documents, Government Printing Office, Washington, D. C. 20402, GPO #1973-546-312/140, $1.50.

Periodicals

ECONOMIC PRIORITIES REPORT: PAPER PROFITS, POLLUTION AUDIT 1972, Vol. 3, No. 3, July/August, 1972. Available from the Council on Economic Priorities, 84 Fifth Avenue, New York, N. Y. 10011, $3.00.

ENVIRONMENTAL LAW REPORTER, Environmental Law Institute, 1346 Connecticut Avenue, NW, Washington, D. C. 20036, $135/yr. Also available at law libraries.

ENVIRONMENTAL REPORTER, Bureau of National Affairs, 1231 25th Street, Washington, D. C. 20037, $296/yr. Also available at law libraries.

FEDERAL REGISTER, daily. Available from Superintendent of Documents, Government Printing Office, Washington, D. C. 20402, include date and title, $.75 each.

Indexes

MOODY'S BOND RECORD, Moody's Investors Service, Inc. Available at most business libraries.

Data Sources

ENVIRONMENTAL IMPACT STATEMENTS, available from agency which prepared it, or National Technical Information Service, 5282 Port Royal Rd., Springfield, Va. 22151 (see Chapter 6).

FORM 67, available at the Federal Power Commission, Washington, D.C.

EMISSIONS INVENTORIES, available at state air pollution control agencies.

NATIONAL EMISSIONS REPORTS, available from the Data Processing Center, Monitoring and Analysis Division, U. S. Environmental Protection Agency, Durham, North Carolina.

NATIONAL POLLUTION DISCHARGE ELIMINATION SYSTEM APPLICATIONS, available at U. S. Environmental Protection Agency regional offices.

OSHA INSPECTION REPORTS, available from Occupational Safety and Health Administration area offices.

REFUSE ACT PERMIT PROGRAM APPLICATIONS, available from U. S. Environmental Protection Agency regional offices.

WATER RESOURCES DATA FOR (STATE), PART I, SURFACE WATER RECORDS, available from Water Resources Division, United States Geological Survey, U. S. Department of the Interior in (State), free.

CHAPTER 5

Books

CITIZEN SUITS UNDER THE CLEAN AIR ACT AMENDMENTS OF 1970, Richard E. Ayres and James F. Miller, Natural Resources Defense Council Newsletter, Vol. 1, No. 4, free.

CITIZENS MAKE A DIFFERENCE: CASE STUDIES OF ENVIRONMENTAL ACTION, Citizens' Advisory Committee on Environmental Quality, 1973. Available from Superintendent of Documents, Government Printing Office, Washington, D. C. 20402, GPO #4000-00290, $1.75.

CLAIRTON TASK FORCE REPORT, Group Against Smog and Pollution, 1972. Available from GASP, P. O. Box 5165, Pittsburgh, Pa. 15206, $3.50.

DON'T LEAVE IT ALL TO THE EXPERTS: THE CITIZEN'S ROLE IN ENVIRONMENTAL DECISION MAKING, U. S. Environmental Protection Agency, 1972. Available from Superintendent of Documents, Government Printing Office, Washington, D. C. 20402, GPO #1972-0-478-748, $.55.

ENVIRONMENTAL LAW, Arnold W. Reitze, Jr., North American International Publishing Company, Washington, D. C., 1972. Out of print, available at libraries.

MINDING THE CORPORATE CONSCIENCE, CEP, 84 Fifth Ave., NYC, $1.

NEWS FOR INVESTORS, IRRC, 1522 K St., Washington, D.C., $48/yr.

POLITICAL HANDBOOK, Sierra Club, 2nd edition, 1973. Available from Sierra Club, 220 Bush Street, San Francisco, California 94104, $.50.

YOUR RIGHT TO CLEAN AIR: A MANUAL FOR CITIZEN ACTION, The Conservation Foundation, 1970. Available from the Conservation Foundation, 1717 Massachusetts Avenue, NW, Washington, D. C. 20036, free.

CHAPTER 6

Books

BACKGROUND INFORMATION FOR PROPOSED NEW SOURCE PERFORMANCE STANDARDS, U. S. Environmental Protection Agency. Available from the National Technical Information Service or the U. S. Environmental Protection Agency, order # varies (see list Chapter 2), 0—$4.50.

DEVELOPMENT DOCUMENTS FOR EFFLUENT LIMITATION GUIDELINES AND NEW SOURCE PERFORMANCE STANDARDS, U. S. Environmental Protection Agency. Available from Superintendent of Documents, Government Printing Office, Washington, D. C. 20402, GPO # varies (see list Chapter 2), $1.60-$20.30.

ENERGY CONSERVATION PROGRAM GUIDE FOR INDUSTRY AND COMMERCE, National Bureau of Standards, 1974. Available from Superintendent of Documents, Government Printing Office, Washington, D. C. 20402, GPO #C13.11:115, $2.50.

ENVIRONMENTAL QUALITY—1974, The President's Council on Environmental Quality, Fifth Annual Report. Available from Superintendent of Documents, Government Printing Office, Washington, D. C. 20402, GPO #4000-00327, $5.20.

INDUSTRIAL DEVELOPMENTS AND THE ENVIRONMENT: LEGAL REFORMS TO IMPROVE THE DECISION-MAKING PROCESS IN INDUSTRIAL SITE SELECTION—REVIEW DRAFT, American Bar Association, 1973. Available from the American Bar Association, 1155 East 60th Street, Chicago, Illinois 60637, free.

THE NATIONAL ENVIRONMENTAL POLICY ACT IN THE COURTS, Prof. Harold Green, The Conservation Foundation, 1972. Available from the Conservation Foundation, 1717 Massachusetts Avenue, NW, Washington, D. C. 20036, $1.00 prepaid.

POTENTIAL FUEL EFFECTIVENESS IN INDUSTRY, Elias Gyftopoulos, Thermo Electron Corporation, 1974. Available from Ballinger Publishing Company, c/o J. B. Lippincott, East Washington Square, Philadelphia, Pennsylvania, paper—$3.00, hard bound—$8.00.

PRICE OF POWER: ELECTRIC UTILITIES AND THE ENVIRONMENT, Charles Komanoff, Holly Miller, Sandy Noyes, Council on Economic Priorities, 1972. Available from the Council on Economic Priorities, 84 Fifth Avenue, New York, New York 10011, $18.50.

PROCESSES, PROCEDURES, AND METHODS TO CONTROL POLLUTION RESULTING FROM ALL CONSTRUCTION ACTIVITY, U. S. Environmental Protection Agency, 1973. Available from Superintendent of Documents, Government Printing Office, Washington, D. C. 20402, EPA #430/9-73-007, $2.30.

STEAM-ELECTRIC AIR AND WATER QUALITY CONTROL DATA, Federal Power Commission, annual. Available from the Federal Power Commission, 825 North Capitol Street, Washington, D. C. 20426, $2.00.

Periodicals

FEDERAL REGISTER, daily. Available from Superintendent of Documents, Government Printing Office, Washington, D. C. 20402, include date and title, $.75 each.

102 MONITOR, monthly, Council on Environmental Quality. Available from Subscription Department, Government Printing Office, Washington, D. C. 20402, GPO #196-62331, $1.80/copy, $21.50/yr.

Data Sources

APPLICATIONS FOR CONSTRUCTION PERMITS, filed with state air pollution control agency or state water pollution control agency.

ENVIRONMENTAL IMPACT STATEMENT, available from agency preparing it or the National Technical Information Service, 5282 Port Royal Rd., Springfield, Va. 22151 (see Chapter 6 for details).

NATIONAL POLLUTION DISCHARGE ELIMINATION SYSTEM PERMIT APPLICATION, filed with the U. S. Environmental Protection Agency regional office and/or the state water pollution agency.

NEW SOURCE REVIEW, available from the U. S. Environmental Protection Agency regional office.

TOPOGRAPHIC MAPS, available from United States Geological Survey, Denver, Colorado, $.75 each.

Glossary

Activated sludge: a sophisticated secondary water pollution treatment process, in which a sludge, rich in bacteria, is whipped into waste water, and the bacteria rapidly consume organic waste.

Aeration lagoon: a basin in which secondary water pollution treatment takes place; aerators force oxygen into the waste water, feeding the bacteria which consume organic wastes.

Air pollution: contaminants in the air in concentrations that interfere directly or indirectly with man's health, safety, comfort and the full use and enjoyment of property, or have adverse effects on plant or animal life.

Air quality criteria: the documented concentration levels and lengths of exposure at which, based on currently available scientific information, specific air pollutants have detectable adverse effects on health and welfare.

Ambient air: the atmosphere; outdoor air.

Ambient air quality standard: a limit on the concentration of a given pollutant permitted in the ambient air.

> **Long-term standards:** as above, typically for one year and expressed as an annual average, often as annual geometric or arithmetic mean.

> **Short-term standards:** as above, typically for periods such as one day, one hour, five minutes, etc.

Asbestos: a fibrous carcinogenic mineral substance.

Baghouse filter: an air pollution control device which uses filters to capture up to 99.9 percent of particulate matter. The structure in which large groups of filters are contained is known as a baghouse.

Basic oxygen furnace: a type of steelmaking furnace.

BAT: Best Available Technology; level of control (also known as "Level II") mandated for all factories for 1983 by the 1972 Federal Water Pollution Control Act.

Biochemical oxygen demand: the amount of oxygen consumed by biological processes to break down a given quantity of organic matter in water; a measure of the organic pollutant load.

BPT: Best Practicable Technology; level of control (also known as "Level I") mandated for all factories for 1977 by the 1972 Water Pollution Control Act.

Btu: British thermal unit, a measure of heat energy.

Carbon monoxide: a colorless, odorless, toxic gas produced by any process that involves the incomplete combustion of carbon substances. Primarily emitted from automobiles.

Carcinogenic: cancer-producing.

Clarifier: a water pollution control device consisting of a basin that provides primary treatment, *i.e.* removal by settling or skimming of waste materials.

Combustion: burning.

Cooling pond: a simple type of thermal pollution control system consisting of a man-made lake in which heated water from a plant is cooled, primarily by evaporation.

Cooling tower: a more advanced thermal pollution control device. Heated water is cooled when circulated through a large tower.

Criteria pollutants: six air pollutants—particulates, sulfur dioxide, hydrocarbons, nitrogen oxides, carbon monoxide, and photochemical oxidants—for which the U.S. EPA has set primary and secondary ambient pollution standards under the

Clean Air Act of 1970. The first five are emitted from industrial operations.

Cyclone collector: an air pollution control device that uses mechanical means to collect particulates. Also known as a mechanical collector.

Dissolved solids: pollutants, primarily salts, carried by water in a dissolved state.

Effluent: waste water.

EIS: Environmental Impact Statement.

Electric furnace: a type of steel-making furnace.

Electrostatic precipitator: an air pollution control device that uses an electric field to trap up to 99.5 percent of particulates in the gas stream.

Emission factor: amount of air pollution generated per unit of output by an uncontrolled industrial process.

Emission limitation (also stack gas or emission) standard: the maximum amount of a pollutant that is permitted to be discharged from a single polluting source, *i.e.* from a smokestack.

Environmental impact statement: a document prepared by a government agency or private entity on the impact of an action significantly affecting the quality of the human environment.

Flue gas: the air emitted to the atmosphere through a smokestack after a production process or combustion takes place; also called stack gas.

Fly ash: all solids, including ash, charred paper, cinders, dust, soot, or other partially incinerated matter, that are carried in a gas stream.

Fossil fuels: coal, oil, and natural gas; so called because they are derived from the remains of ancient plant and animal life.

Gas stream: the air, clean or polluted, that is present during a production process or combustion, and is eventually vented to the atmosphere.

Hydrocarbon: a vast family of compounds containing carbon and hydrogen in various combinations; found especially in fossil fuels. Some hydrocarbon compounds are major air pollutants and may be carcinogenic.

Intermittent controls: a method of reducing air pollution concentrations by shutting down production processes when pollution rises above dangerous levels.

Lime kiln: a piece of production equipment at a kraft pulp mill; it processes chemicals used in making wood into pulp.

Limestone scrubber: an air pollution control device which uses limestone in removing sulfur dioxide from smokestack gases.

Mechanical collector: one of the oldest and least efficient air pollution control devices; it collects heavy particulates from the gas stream.

Monitoring: the sampling of air and water for measurement of the amount of pollutants in the environment.

Nitrogen oxides: gases formed from atmospheric nitrogen and oxygen when combustion takes place under conditions of high temperature and pressure, *e.g.* in auto engines and power plants.

NSPS: New Source Performance Standard, established by the Federal EPA for particular industries.

Oxide: a compound of two elements, one of which is oxygen.

Particulate: a particle of solid or liquid matter suspended in the air; soot and dust.

pH: a measure of acidity and alkalinity of water, generally on a scale of 1 (most acid), through 7 (neutral), to 14 (most alkaline).

Point source: a stationary source of pollution, generally an industrial smokestack.

Power boiler: a piece of equipment in which fuels are burned, producing steam to run industrial processes.

Primary standard: a limit on the concentration of a criteria pollutant allowable in ambient air; defined by the Clean Air Act as a level stringent enough to protect the public health. Established by the U.S. EPA.

Primary treatment: the first stage in waste water treatment, in which floating or settleable solids are mechanically removed by skimming or settling.

Raw waste load: the amount of water pollution generated per unit of output by an uncontrolled industrial process.

Recovery boiler: a piece of production equipment at a kraft pulp mill; it processes chemicals used in turning wood into pulp.

Recycling: the process of transforming wastes into new products.

Sanitary landfill: a site for solid waste disposal at which wastes are buried under layers of dirt.

Secondary standard: a limit on the concentration of a criteria pollutant allowable in ambient air; defined by the Clean Air Act as a level stringent enough to protect the public welfare and vegetation. Established by the U.S. EPA.

Secondary treatment: biological treatment; the second step in water pollution control in which bacteria consume organic wastes. Usually accomplished in an aeration lagoon.

Sludge: material removed from waste water by treatment systems.

Smog: the irritating brown haze resulting from the sun's effect on certain pollutants in the air, notably hydrocarbons and nitrogen oxides.

Solid waste: useless, unwanted, or discarded materials with insufficient liquid content to be free flowing.

State-of-the-art control: the best commercially available pollution control system.

Sulfur dioxide: a heavy and irritating colorless gas formed primarily by combustion of coal, oil, and other sulfur-bearing fuels. Also produced in chemical plants and metal smelters.

Suspended solids: small particles of solid pollutants suspended in water.

Tertiary treatment: a catch-all phrase for anything beyond secondary waste water treatment, involving the use of chemical methods and advanced physical techniques.

Thermal pollution: the degradation of water quality by the introduction of a heated effluent.

Thermal pollution control: the cooling off of heated water prior to discharge; accomplished by use of cooling ponds or cooling towers.

Toxicity: the degree of being poisonous or harmful to plant or animal life.

Trickling filter: a water pollution control device that provides secondary treatment.

Variance: an agreement by a governing body to a delay or exception in the application of a given law, ordinance, or regulation.

Water pollution: sewage, industrial wastes, or other harmful or objectionable material in water in sufficient concentrations or quantities to result in measurable degradation of water quality.

Water quality criteria: the levels of pollution that affect the suitability of water for a given use.

Water quality standard: a limit on the amount of water pollution permitted in a body of water so that it will still be clean enough for a specified use (recreation, fish and wildlife propagation, drinking water, industrial or agricultural).

Wet scrubber: an air pollution control device that uses a liquid spray to remove aerosol, gaseous, and particulate pollutants from an air stream.

PHOTO CREDITS

Frontpiece	EPA Documerica, Marc St. Gil
page 11	International Paper Company
page 18 top	EPA Documerica, Cornelius Keyes
bottom	Stephen Moody
page 21	Mobil Oil Corporation
page 29	EPA Documerica, Marc St. Gil
page 40	International Paper Company
page 44	International Paper Company
page 46	EPA Documerica, Gene Daniels
page 52	U.S. Environmental Protection Agency
page 65	EPA Documerica, LeRoy Woodson
page 73	U.S. Environmental Protection Agency
page 84	U.S. Environmental Protection Agency
page 106	Atomic Industrial Forum, Inc.
page 144	EPA Documerica, Cornelius Keyes
page 151 top	EPA Documerica
bottom	Charles P. Noyes
page 159	EPA Documerica, John Alexandrowicz
page 164	Group Against Smog & Pollution (GASP)
page 174	EPA Documerica, Joe Clark
page 178	International Paper Company
page 189	EPA Documerica, Charles O'Rear
page 199	American Oil Company
page 205	Mobil Oil Corporation
page 213 top	EPA Documerica, Charles O'Rear
bottom	EPA Documerica, Lyntha Scott Eiler

ndex

Afterword

U.S. corporations produce and sell most of the goods and services used in this country. In the process, they determine the nature of work and working conditions for over 40 million people (almost half the total work force); they control one third of the nation's tangible wealth; they make major decisions as to the consumption of natural resources and the generation of air and water pollution; they establish needs and requirements for many public services and substantively shape American values and buying patterns. The corporation, while functioning as an economic entity, thus plays a significant role in forming public policy.

INFORM, a non-profit, tax-exempt organization, was set up in 1973 to study the social role and impact of U.S. industries, particularly on the environment, on employees, and on consumers. The organization's focus is not only on assessing the extent of corporate social impacts but also on defining the kinds of programs and practices available to industries to improve their performance (and the costs of such programs).

INFORM's research is published in books and reports. These publications should help clarify for concerned Americans— workers, government officials, business managers, investors, public interest groups—some of today's most serious corporate problems and the options for change. The publications will also attempt to provide citizens with tools they can use to study and evaluate industry practices themselves. A CLEAR VIEW, which we are pleased to release as INFORM's first book, exemplifies the type of educational tool we hope to offer, enabling people to conduct their own research in a more knowledgeable and efficient manner. (The four other studies now under way at INFORM focus on land use, energy and occupational safety and health issues.)

INFORM has been largely supported during its first two years by grants from a number of foundations: Rockefeller, Abelard, Mary Reynolds Babcock, Shalan, Spoonbill, Norman, New York and Faigel Leah. Our work is now increasingly funded by individual and institutional subscribers, who receive books, abstracts of report findings, and the quarterly *INFORM News* which discusses projects and progress of the organization.

Brief descriptions of all books in preparation at INFORM now, a listing of our staff, Steering Committee and Advisory Board members, and annual subscription rates to INFORM research follow.

INFORM STUDIES IN PREPARATION

An analysis of the effects of the U.S. land subdivision industry on consumers and the environment: how corporations buy, subdivide, and sell large tracts of rural land, how development operations alter site ecology, how government regulates these activities.

An analysis of corporate research and development of new and more environmentally sound energy sources (solar, geothermal, fuel cell, etc.).

An analysis of occupational health and safety problems at mining and smelting operations in the non-ferrous metals industry, and the adequacy of governmental and corporate efforts to control them.

A study of corporate involvement in urban preservation work: profiles of projects aimed at restoration and reuse of the man-made environment; a review of business and governmental incentives to encourage conservation activities.

INFORM STAFF AND ADVISORS

INFORM Staff: Director, Joanna Underwood; Editor, Jean Halloran; Research Associates, Leslie Allan, James Cannon, Stewart W. Herman, Beryl Kuder, Sarah Oakes, Mary Roman; Research Assistant, Joel Katzman; Administrator, Pat Konecky; Administrative Assistant, Ilene Steingut; Marketing, Elizabeth Buckner, Elizabeth Vagliano.

Steering Committee: Robert Alexander, Consultant, McKinsey & Co.; Marshall Beil, Lawyer, Karpatkin, Pollet & LeMoult; Susan Butler, Staff Associate, Planned Parenthood of New York; William Butler, Marketing Consultant; Albert Butzel, Partner, Berle, Butzel & Kass; Charles Costello, Office of General Council, AID; Elizabeth Durbin, Prof. of Economics, New York University; Timothy Hogan, Program Mgr., SCA Services, Inc; Charles Komanoff, Energy Consultant; Lewis Kruger, Partner, Krause, Hirsch & Gross; Raymond Maurice, Sociologist; John P. Milton, Chairman, Threshold, International Center for Environmental Research; Charles P. Noyes, Environmental writer and photographer; Herschel E. Post, Asst. Vice President, Morgan Guaranty Trust Co.; Martha Stuart, President, Communications for Change; Edward H. Tuck, Partner, Shearman and Sterling; Ranne Warner, Research Associate, Harvard Business School; Anthony Wolff, Environmental Writer.

Advisory Council: Fred S. Dubin, President, Dubin, Mindell, Bloome Associates, Consulting Engineers; Herbert J. Gans, Prof. of Sociology, Columbia University; LaDonna Harris, President, Americans for Indian Opportunity; Hon. Paul N. McCloskey, Jr., Congressman, Cal.; Stewart Udall, Chairman, Overview.